To Dr. Jianhong Wu, with gratitude for your guidance and support.

Contents

Preface

Data clustering, an interdisciplinary field with diverse applications, has gained increasing popularity since its origins in the 1950s. Over the past six decades, researchers from various fields have proposed numerous clustering algorithms. In 2011, I wrote a book on implementing clustering algorithms in C++ using object-oriented design and programming techniques [94]. While C++ offers efficiency and speed, it has a steep learning curve, making it less ideal for prototyping clustering algorithms.

Fourteen years have passed since the publication of that book, and much has changed in the machine learning and data mining communities. Notably, Python has surged in popularity. According to the TIOBE Index [1], Python has been the most widely used programming language since 2022. Its clear syntax and extensive ecosystem of libraries have made it easier than ever to implement clustering algorithms efficiently.

This book extends the concepts from my C++ book by implementing clustering algorithms in Python. While some chapters overlap, the implementation approach differs significantly. The C++ book adopted an object-oriented framework to maximize code reuse. In contrast, this book employs a procedural programming approach, as many clustering algorithms can be implemented in Python with just a few lines of code.

The book is divided into two parts. The first part introduces Python and key libraries commonly used in scientific computing and data science, including NumPy, Pandas, and Matplotlib. The second part focuses on implementing popular clustering algorithms, covering both hierarchical and partitional methods.

Each chapter on clustering algorithms follows a structured format. I begin with a theoretical overview, followed by the Python implementation, and conclude with demonstrations using both synthetic and real-world data. For some algorithms, I also compare our implementations with those in the scikit-learn package.

This book is suitable for anyone interested in learning data clustering and implementing clustering algorithms in Python. No prior knowledge of Python is required. Readers already familiar with Python may skip the first part and proceed directly to the second. Most of the Python code is included within the book, and the complete set of source files is available at:

https://github.com/ganml/dcpython

---

[1] https://www.tiobe.com/tiobe-index/. Accessed on February 1, 2025.

Hands-on practice is one of the best ways to learn. I encourage readers to experiment with the code and examples, modify the implementations, and explore new clustering algorithms. Unlike the implementations in well-known Python libraries such as scikit-learn, the code in this book is designed for clarity and simplicity, assuming users will provide valid input without extensive error handling.

Finally, I would like to acknowledge the support provided by the Makuch Faculty Fund from the Department of Mathematics at the University of Connecticut.

Storrs, CT
Guojun Gan

Part I

Python Programming
Preliminaries

1

Python Programming 101

Python is one of the most widely used programming languages today. In this chapter, we provide a concise introduction to the Python programming language, covering essential topics for beginners. Specifically, we will discuss how to set up Python environments, explore the basics of Python programming, and introduce some common packages frequently used in data science.

1.1 Installation

A straightforward way to install Python on your computer is to use the individual edition of the Anaconda distribution, available at `https://www.anaconda.com`. This edition is open-source and compatible with Windows, macOS, and Linux. Anaconda offers two versions: the full Anaconda distribution and Miniconda. The full distribution includes over 300 pre-installed data science and machine learning packages but requires more disk space. In contrast, Miniconda is a lightweight installer that includes only conda, Python, and a few essential packages.

In this book, we will use the Miniconda distribution since we do not require all the packages included in the full Anaconda distribution. The latest Miniconda installer can be downloaded from `https://docs.anaconda.com/miniconda/`. At the time of writing this book, the most recent version of Miniconda was Conda 24.7.1, released on August 22, 2024, for Python 3.12.4.

Once the Miniconda is installed, we can open the Anaconda Prompt. In a Windows computer, for example, typing `dir` in the Anaconda Prompt gives the following output:

```
1  (base) C:\Users\gjgan>dir
2  Directory of C:\Users\gjgan
3
4  09/08/2024  10:14 AM    <DIR>          .
5  07/10/2024  02:43 PM    <DIR>          ..
6  07/17/2024  09:53 PM    <DIR>          .cache
7  09/08/2024  10:14 AM    <DIR>          .conda
8  07/10/2024  08:09 PM    <DIR>          .eclipse
9  07/18/2024  08:53 AM                54 .gitconfig
```

DOI: 10.1201/9781003592648-1   3

```
10  07/10/2024    03:02 PM    <DIR>          .m2
11  07/10/2024    06:36 PM    <DIR>          .ms-ad
12  07/11/2024    02:21 PM    <DIR>          .openjfx
13  09/06/2024    07:32 AM    <DIR>          .p2
14  07/10/2024    02:42 PM    <DIR>          .ssh
15  07/14/2024    04:04 PM    <DIR>          Canon
16  07/10/2024    02:15 PM    <DIR>          Contacts
17  07/10/2024    04:14 PM    <DIR>          Documents
18  09/08/2024    12:52 PM    <DIR>          Downloads
19  07/10/2024    02:44 PM    <DIR>          eclipse
20  07/10/2024    02:15 PM    <DIR>          Favorites
21  07/10/2024    02:15 PM    <DIR>          Links
22  09/08/2024    10:14 AM    <DIR>          miniconda3
23  07/10/2024    02:15 PM    <DIR>          Music
24  09/05/2024    11:25 PM    <DIR>          OneDrive
25  07/10/2024    02:15 PM    <DIR>          Saved Games
26  07/10/2024    03:21 PM    <DIR>          Searches
27  07/11/2024    10:09 PM    <DIR>          Videos
28  1 File(s)                 54 bytes
29  23 Dir(s)    922,031,411,200 bytes free
30
31  (base) C:\Users\gjgan>
```

From the above output, we see the directory miniconda3, which is the installation location for Miniconda.

Once Miniconda is installed, we can create a conda environment, which is a self-contained, isolated space where specific versions of Python packages can be installed. Using a conda environment is highly beneficial as it helps avoid conflicts between Python packages. To create a conda environment, we can define the required Python packages in a YAML (Yet Another Markup Language) file and then use the Conda Prompt to build the environment.

```
1   name: dc
2   channels:
3     - conda-forge
4     - pytorch
5   dependencies:
6     - cython
7     - python
8     - pytorch
9     - numpy
10    - pandas
11    - matplotlib
12    - scikit-learn
13    - ucimlrepo
14    - spyder
```

Listing 1.1: Content of the YAML file dc.yml used to create the conda environment.

Listing 1.1 shows the content of a YAML file that can be used to create a conda environment. Suppose that the YAML file was saved to the directory documents. To create the conda environment, we execute the following command in the Conda Prompt:

```
conda env create -f documents\dc.yml
```

It may take a few minutes for the command to complete. Once the command finishes executing, the following output will appear in the Conda Prompt:

```
Channels:
  - conda-forge
  - pytorch
  - defaults
Platform: win-64
Collecting package metadata (repodata.json): done
Solving environment: done

Downloading and Extracting Packages:

Preparing transaction: done
Verifying transaction: done
Executing transaction: done
#
# To activate this environment, use
#
#     $ conda activate dc
#
# To deactivate an active environment, use
#
#     $ conda deactivate
```

The conda environment created using the YAML file includes several common Python packages: NumPy, Pandas, Matplotlib, scikit-learn, and uciml-repo. Additionally, the environment comes with Spyder, a Python code editor. Spyder is an open-source integrated development environment (IDE) for Python that supports advanced editing, interactive testing, and debugging. Figure 1.1 shows a screenshot of the Spyder interface when the program is launched. To check the versions of the installed packages, execute the following command in the Conda Prompt:

```
conda list -c "^(numpy|pandas|matplotlib|scikit|ucimlrepo|
    spyder)"
```

After executing the command in the Conda Prompt, the following information was displayed:

```
conda-forge/win-64::matplotlib-3.10.0-py312h2e8e312_0
conda-forge/win-64::matplotlib-base-3.10.0-py312h90004f6_0
conda-forge/noarch::matplotlib-inline-0.1.7-pyhd8ed1ab_1
conda-forge/win-64::numpy-2.2.1-py312hf10105a_0
```

```
 5  conda-forge/noarch::numpydoc-1.8.0-pyhd8ed1ab_1
 6  conda-forge/win-64::pandas-2.2.3-py312h72972c8_1
 7  conda-forge/win-64::scikit-learn-1.6.0-py312h816cc57_0
 8  conda-forge/win-64::spyder-6.0.3-py312h2e8e312_1
 9  conda-forge/noarch::spyder-kernels-3.0.2-win_pyh7428d3b_0
10  conda-forge/noarch::ucimlrepo-0.0.7-pyhd8ed1ab_0
```

Since Python packages are updated frequently, the versions you see when building the conda environment may differ from those listed above.

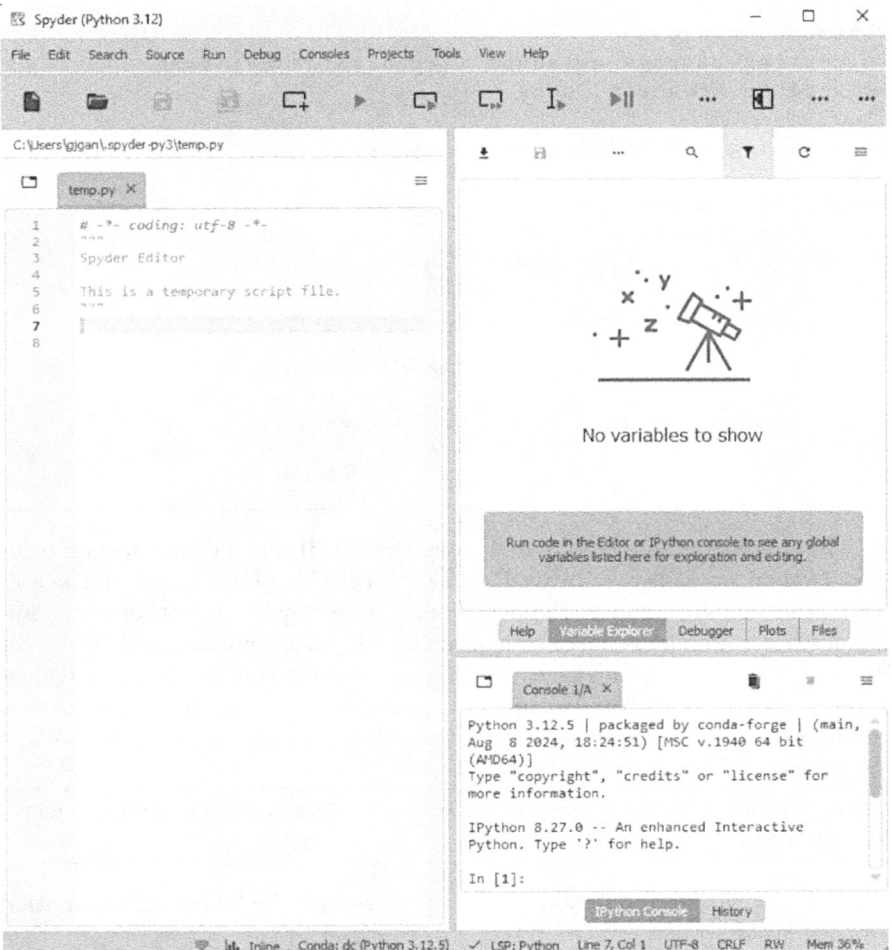

FIGURE 1.1: A screenshot of the Spyder editor.

1.2   Variables and data types

In programming, a variable is a named storage location that holds a value or data. Variables allow us to reference and manipulate the stored values or data. Unlike in languages such as Java or C++, Python does not require explicit variable declarations. Instead, a variable is created simply by assigning a value to it:

```
pi = 3.1415926
print(pi)
```

In the code above, we create a variable named `pi` and assign it the value 3.1415926. We then use Python's `print` function to display the value stored in the variable `pi`. When the code is executed, the output 3.1415926 is displayed.

Python supports multiple assignments in a single line. For example, we can assign values to two variables simultaneously, as shown below:

```
x, y = 1, 2
```

The line of code above assigns the value 1 to the variable x and 2 to the variable y. Multiple assignments can also be used to swap the values of variables. For example, in the following code, we first assign values to three variables, x, y, and z, and then swap their values:

```
x, y, z = 1, 2, 3
print(x)
print(y)
print(z)
x, y, z = y, z, x
print(x)
print(y)
print(z)
```

When the above block of code is executed in Spyder, the following output is displayed:

```
1
2
3
2
3
1
```

In Python, variable names must adhere to the following rules:

- Valid characters include lowercase letters, uppercase letters, numbers, and underscores (_).

- A variable name must begin with a letter or an underscore and cannot start with a number.

- A variable name cannot be a Python keyword (reserved word).

To view a list of Python keywords, we can run the following code:

```
import keyword
print(keyword.kwlist)
```

After running the code above, the following output was displayed:

```
['False', 'None', 'True', 'and', 'as', 'assert', 'async', '
    await', 'break', 'class', 'continue', 'def', 'del', '
    elif', 'else', 'except', 'finally', 'for', 'from', '
    global', 'if', 'import', 'in', 'is', 'lambda', '
    nonlocal', 'not', 'or', 'pass', 'raise', 'return', 'try
    ', 'while', 'with', 'yield']
```

The current version of Python includes 35 keywords, which cannot be used as variable names. Additionally, variable names in Python are case-sensitive. For example, `pi` and `Pi` are considered distinct variable names.

Python has four fundamental data types: integers, floats, booleans, and strings. An integer represents a whole number without a decimal point, while a float represents a real number with decimal precision. A boolean is a binary type that can only take the values `True` or `False`. A string is a sequence of characters enclosed in either single or double quotes.

In the following code, we create four variables, each assigned a value of a different data type. We then use Python's `type` function to determine the data type of each variable.

```
bA = True
iB = 2024
fC = 3.14
sD = 'data'
type(bA)
type(iB)
type(fC)
type(sD)
```

If we run the code line by line in Spyder, the following output will be displayed:

```
bA = True

iB = 2024

fC = 3.14

sD = 'data'
```

```
 8
 9 type(bA)
10 Out[1148]: bool
11
12 type(iB)
13 Out[1149]: int
14
15 type(fC)
16 Out[1150]: float
17
18 type(sD)
19 Out[1151]: str
```

Strings are a special type of data. A string is essentially a sequence of characters, where each character has a specific position, known as its index. When indexed from left to right, the characters are assigned indices starting from 0 up to $n - 1$, where n is the length of the string. When indexed from right to left, the characters are assigned indices starting from -1 down to $-n$. In the following code, we define a string, determine its length, and extract substrings using indexing:

```
1 s1 = "a string"
2 len(s1)
3 s1[0:2]
4 s1[-1]
5 s1[-3:-1]
6 s1[-3:]
```

When the above code is executed line by line in Spyder, the following output will be displayed:

```
 1 s1 = "a string"
 2
 3 len(s1)
 4 Out[34]: 8
 5
 6 s1[0:2]
 7 Out[35]: 'a '
 8
 9 s1[-1]
10 Out[36]: 'g'
11
12 s1[-3:-1]
13 Out[37]: 'in'
14
15 s1[-3:]
16 Out[38]: 'ing'
```

TABLE 1.1: A list of operators and functions for manipulating strings

Function	Description
+	Concatenates two strings.
*	Repeats a string a specified number of times.
[:]	Slices a string based on character positions.
in	Checks if a string is a substring of another string.
replace	Replaces a substring within a string with another substring.
split	Splits a string into a list of substrings using a specified delimiter.
format	Formats a string by replacing placeholders () with specified values.
join	Joins a list of strings into a single string using a specified delimiter.

Table 1.1 lists some commonly used operators and functions for manipulating strings. The following block of code demonstrates how to use these operators and functions:

```
print("a " + "string")
print("a" * 5)
print("string"[0:3])
print("str" in "string")
print("string".replace("s", "S"))
print("1,2,3".split(","))
print("{} and {} are the results".format(1, 2))
print(",".join(["1", "2", "3"]))
```

When the above block of code is executed in Spyder, the following output is displayed:

```
a string
aaaaa
str
True
String
['1', '2', '3']
1 and 2 are the results
1,2,3
```

1.3 Data structures

Python has four fundamental data structures: lists, tuples, dictionaries, and sets. A list is an ordered collection of elements, which can be of any data type. Lists are defined using square brackets [], with elements separated by commas. The following code demonstrates a list containing three items:

```
lst1 = ["a", 1, True]
print(lst1)
print(lst1[0])
```

The three elements have different data types. When the above block of code is executed in Spyder, the following output is displayed:

```
['a', 1, True]
a
```

Lists are mutable. For example, we can add an element to the end of a list by using the function **append**:

```
lst1.append(3.14)
print(lst1)
```

In the above code, we add the number 3.14 to the end of the list `lst1`. Executing the above block of code, we get the following output:

```
['a', 1, True, 3.14]
```

We can also insert an element into a specified position of a list. In the following code, we insert the number 3.14 before the position 0:

```
lst1.insert(0, 3.14)
print(lst1)
```

The output of the above code is

```
[3.14, 'a', 1, True, 3.14]
```

Elements of a list can be accessed by indexing and slicing. Table 1.2 shows different methods for accessing elements of a list. The indices of elements of a list have two forms. For a list with n elements, the indices of the elements are $0, 1, \ldots, n-1$ or $-n, -(n-1), \ldots, -1$.

The following code gives some examples of accessing elements of `lst1`:

```
print(lst1[0])   # get the first element
print(lst1[-1])  # get the last element
print(lst1[1:3]) # get the second and the third elemnts
print(lst1[-3:]) # get the last three elements
```

TABLE 1.2: Methods for accessing elements in a list.

Method	Description
[pos]	Gets the element with index pos
[start:stop]	Gets elements with indices from start (inclusive) to stop (exclusive)
[start:stop:stride]	Gets elements with indices from start (inclusive) to stop (exclusive) with step size stride

```
5  print(lst1[1:]) # get all elements except the first one
6  print(lst1[:3]) # get the first three elements
7  print(lst1[::-1]) # get elements with indices -1, -2, ...
8  print(lst1[0:3:2]) # get elements 0, 2
9  print(lst1[-2:-4:-2]) # get element -2
```

The pound sign # in the above code is the comment symbol in Python. Code appears after the pound sign will not be executed. Executing the above code gives the following output:

```
1  3.14
2  3.14
3  ['a', 1]
4  [1, True, 3.14]
5  ['a', 1, True, 3.14]
6  [3.14, 'a', 1]
7  [3.14, True, 1, 'a', 3.14]
8  [3.14, 1]
9  [True]
```

From the output, we see that the method [::-1] essentially reverses the list.

We can change values of individual elements in a list by assigning new values to them. To change the first two elements, we can proceed as follows:

```
1  lst1[0:2] = [-1, -2]
2  print(lst1)
```

In the above code, we change the first two elements to be -1 and -2, respectively. The output of the above code is

```
1  [-1, -2, 1, True, 3.14]
```

Table 1.3 lists some commonly used functions for lists.

A tuple is a data structure that is similar to a list. However, tuples are immutable. Once a tuple is created, it cannot be modified. Tuples are defined by parentheses and elements of a tuple are separated by commas. The following code creates two tuples of different sizes:

TABLE 1.3: Python functions for lists.

Function	Description
append	Adds an element to the end of a list
clear	Removes all elements from a list
copy	Creates a copy of a list
count	Counts the number of elements with the specified value
extend	Adds all elements of a collection to the end of a list
index	Gets the index of the first element with the specified value
insert	Inserts an element to the specified position
len	Counts the number of elements in a list
pop	Removes an element at the specified position
remove	Removes the first element with the specified value
reverse	Reverses the order of the elements in a list
sort	Sorts the elements of a list

```
t1 = (1, 3)
t2 = ("a", True, 1)
print(t1)
print(t2)
```

Executing the above code creates two tuples and gives the following output:

```
(1, 3)
('a', True, 1)
```

Elements of a tuple can be accessed by square brackets with indices. The following code shows how to get elements of tuples:

```
print(t1[1])
print(t2[1:3])
```

Executing the above code gives the following output:

```
3
(True, 1)
```

Unlike lists and tuples, a dictionary is a data structure used to store key-value pairs. A dictionary is created by using curly brackets. Key-value pairs are separated by commas. The key and the value of a pair are separated by a colon. The following code creates a dictionary with two elements:

```
m1 = {"a" : 1, "b" : 2}
print(m1)
```

Executing the above code gives the following output:

```
{'a': 1, 'b': 2}
```

Values in a dictionary can be accessed by their keys. For example, the following code shows how to access values stored in m1:

```
print(m1["a"])
print(m1["b"])
```

The output of the above code is

```
1
2
```

TABLE 1.4: A list of functions for dictionaries.

Function	Description
clear	Removes all elements from a dictionary
copy	Creates a copy of a dictionary
fromkeys	Returns a dictionary with the specified keys and a specified value
get	Returns the value of a specified key
items	Returns a view object of all key-value pairs of a dictionary
keys	Returns a view object of all keys of a dictionary
pop	Removes a specified key and return the corresponding value
setdefault	Inserts a key with a specified value if the key does not exist; otherwise, returns the value of the key
update	Updates a dictionary with the specified key-value pairs
values	Returns a view object of all values of a dictionary

Table 1.4 provides a list of functions for manipulating dictionaries. For instance, to retrieve the keys and values of the dictionary m1 created earlier, we can use the following code:

```
print(m1.keys())
print(m1.values())
print(m1.items())
```

Executing the above code gives the following output:

```
dict_keys(['a', 'b'])
dict_values([1, 2])
dict_items([('a', 1), ('b', 2)])
```

The outputs given above are view objects, which dynamically reflect any changes made to the corresponding dictionary. These view objects can be converted into lists using the `list` function. For example, the view objects can be transformed into lists as follows:

```
print(list(m1.keys()))
print(list(m1.values()))
print(list(m1.items()))
```

The output of the above code is

```
['a', 'b']
[1, 2]
[('a', 1), ('b', 2)]
```

A set is a collection of unordered and unique elements. In Python, a set can be created using curly brackets {} or the `set` function. For example, the following code demonstrates how to create two sets:

```
s1 = set()
s1.add(1)
s1.add("a")
s2 = {1, "a", "b", True, False}
print(s1)
print(s2)
```

In this example, `s1` and `s2` are sets containing unique elements. Sets automatically eliminate duplicate values, ensuring that each item appears only once. Additionally, sets are unordered, meaning that the elements are not stored in any specific sequence. Executing the above block of code gives the following output:

```
{1, 'a'}
{False, 1, 'a', 'b'}
```

When we create the set `s2`, the boolean value `True` is included in the curly brackets. However, the boolean value `True` does not appear in the output. The reason is that non-zero values in Python are treated as true.

Table 1.5 provides a list of commonly used functions for working with sets. Among these, two functions are particularly useful for removing elements from a set: `discard` and `remove`. The key difference between them is that `discard` does not raise an error if the set does not contain the specified element, whereas `remove` will raise an error in such cases. This distinction makes `discard` a safer option when the presence of the element is uncertain.

1.4 Operators

Operators are special symbols that are used to perform operations on variables and values. They can be categorized into several types based on their

TABLE 1.5: A list of functions for manipulating sets.

Function	Description
add	Adds an element to a set
clear	Removes all elements from a set
copy	Creates a copy of a set
difference	Returns a set containing the difference between two or more sets
discard	Removes an element from a set
intersection	Returns a set containing the common elements of this set and another set
isdisjoint	Returns whether this set is disjoint to another
issubset	Returns whether this set is a subset of another set
issuperset	Returns whether another set is a subset of this set
pop	Removes an element from a set
remove	Removes a specified element from a set
symmetric_difference	Returns a set containing symmetric differences this set and another set
union	Returns a set containing the union of this set and another set
update	Updates a set with its union with other sets

functionality: arithmetic operators, assignment operators, comparison operators, logical operators, identity operators, membership operators, and bitwise operators.

Arithmetic operators are used to perform basic mathematical operations, including addition, subtraction, multiplication, and division. Table 1.6 provides a list of arithmetic operators supported by Python. The first five operators in Table 1.6 are straightforward and intuitive. The modulus operator and the floor division operator are closely related. The modulus operator returns the remainder when the first operand is divided by the second operand, while the floor division operator returns the largest integer less than or equal to the quotient of the division. Let x and y ($y \neq 0$) be two real numbers. Then we have

$$x = y * (x//y) + (x\%y).$$

The following block of code demonstrates the usage of arithmetic operators:

```python
print("2 + 3 =", 2 + 3)
print("2 - 3 =", 2 - 3)
print("2 * 3 =", 2 * 3)
print("2 / 3 =", 2 / 3)
print("2 ** 3 =", 2 ** 3)
print("3.5 % 0.6 =", 3.5 % 0.6)
print("3.5 // 0.6 =", 3.5 // 0.6)
```

TABLE 1.6: Arithmetic operators in Python.

Operator	Description
+	Addition
−	Subtraction
*	Multiplication
/	Division
**	Exponentiation
%	Modulus
//	Floor division

This code illustrates how each arithmetic operator works in Python, including addition, subtraction, multiplication, division, floor division, modulus, and exponentiation. Executing the above block of code in Spyder gives the following output:

```
2 + 3 = 5
2 - 3 = -1
2 * 3 = 6
2 / 3 = 0.6666666666666666
2 ** 3 = 8
3.5 % 0.6 = 0.5000000000000001
3.5 // 0.6 = 5.0
```

Assignment operators are used to assign values to variables. Table 1.7 lists the assignment operators supported by Python. The most basic assignment operator is =, which assigns a value to a variable. Other assignment operators are compound assignment operators, which combine an arithmetic operator with the assignment operator. These operators perform an arithmetic operation on the two operands and then assign the result to the left operand. For instance, the expression x += 2 is shorthand for x = x + 2, where the value of x is incremented by 2 and the result is stored back in x.

TABLE 1.7: Assignment operators in Python.

Operator	Description
=	Assignment
+=	Addition assignment
−=	Subtraction assignment
*=	Multiplication assignment
/=	Division assignment
**=	Exponentiation assignment
%=	Modulus assignment
//=	Floor division assignment

The following block of code demonstrates the usage of the assignment operators:

```
1  x = 3.5
2  print("x =", x)
3  x += 0.6
4  print("x += 0.6:", x)
5  x -= 0.6
6  print("x -= 0.6:",x)
7  x *= 0.6
8  print("x *= 0.6:",x)
9  x /= 0.6
10 print("x /= 0.6:",x)
11 x %= 0.6
12 print("x %= 0.6:",x)
13 x //= 0.6
14 print("x //= 0.6:",x)
```

Executing the above block of code in Spyder gives the following output:

```
1  x = 3.5
2  x += 0.6:  4.1
3  x -= 0.6:  3.499999999999996
4  x *= 0.6:  2.099999999999996
5  x /= 0.6:  3.499999999999996
6  x %= 0.6:  0.49999999999999967
7  x //= 0.6:  0.0
```

Comparison operators are used to compare two values. Table 1.8 lists the comparison operators supported in Python. The following block of code illustrates these comparison operators:

```
1  print("1 == 2:", 1 == 2)
2  print("1 != 2:", 1 != 2)
3  print("1 < 2:", 1 < 2)
4  print("1 > 2:", 1 > 2)
5  print("1 <= 2:", 1 <= 2)
6  print("1 >= 2:", 1 >= 2)
```

The output of the above block of code is

```
1  1 == 2:  False
2  1 != 2:  True
3  1 < 2:  True
4  1 > 2:  False
5  1 <= 2:  True
6  1 >= 2:  False
```

In many programming languages, including Python, careful consideration is required when using comparison operators due to potential numerical

TABLE 1.8: Comparison operators in Python.

Operator	Description
==	Equal to
!=	Not equal to
<	Less than
>	Greater than
<=	Less than or equal to
>=	Greater than or equal to

precision errors. For instance, the expression 3.5%0.6 yields 0.5. Therefore, the comparison 3.5%0.6 > 0.5 should evaluate to false, as the result of the modulus operation is not greater than 0.5. Consider the following piece of code

```
print("3.5 % 0.6 > 0.5:", 3.5 % 0.6 > 0.5)
```

Executing the above code gives the following output

```
3.5 % 0.6 > 0.5: True
```

This is caused by numerical precision errors. To fix this problem, we need to add a small number in the comparison. For example, we can fix the above problem as follows:

```
print("3.5 % 0.6 > 0.5:", 3.5 % 0.6 > 0.5 + 1e-10)
```

The above code will produce the correct answer:

```
3.5 % 0.6 > 0.5: False
```

Logical operators are used to combine or evaluate conditional statements. Table 1.9 lists the primary logical operators supported in Python. These operators allow you to create more complex conditions by connecting multiple expressions, enabling decisions based on combinations of true or false values. The following piece of code illustrates the usage of the logical operators:

```
print("1 < 2 and 1 > 2: ", 1 < 2 and 1 > 2)
print("1 < 2 or 1 > 2: ", 1 < 2 or 1 > 2)
print("not 1 < 2:", not 1 < 2)
```

Executing the above block of code in Spyder gives the following output:

```
1 < 2 and 1 > 2:  False
1 < 2 or 1 > 2:  True
not 1 < 2: False
```

TABLE 1.9: Logical operators in Python.

Operator	Description
and	Returns true if both operands are true
or	Returns true if at least one operand is true
not	Reverses the meaning of the operand

In addition to the operators mentioned earlier, Python also supports other types of operators, such as identity operators and membership operators, which are used to check object identity or membership within a sequence, respectively. Table 1.9 provides a list of these operators. The following piece of code illustrates the usage of these operators:

```
x = 1
y = [1, 2, 3]
z = [1, 2, 3]
print("x is y:", x is y)
print("x is not y:", x is not y)
print("y is z:", y is z)
print("x in y:", x in y)
print("x not in y:", x not in y)
print("y in z", y in z)
```

The output of the above piece of code is:

```
x is y: False
x is not y: True
y is z: False
x in y: True
x not in y: False
y in z False
```

The identity operators, `is` and `is not`, compare the memory locations of two objects. In the example above, we observe that although y and z have identical content, they are not the same object because they reside in different memory locations.

TABLE 1.10: Identity and membership operators in Python.

Operator	Description
is	Return true if the operands are the same
is not	Return true if the operands are different
in	Return true if the second operand contains the first operand
not in	Return true if the second operand does not contain the first operand

Highest

Parentheses ()

Exponentiation **. Right-to-left associativity.

Unary operators +x, -x

Arithematic operators *, /, //, %. Left-to-right associativity

Arithematic operators +, -. Left-to-right associativity

Comparison, identity, membership operators. Left-to-right associativity

Logical operator **not**. Right-to-left associativity

Logical operator **and**. Left-to-right associativity

Logical operator **or**. Left-to-right associativity

Lowest

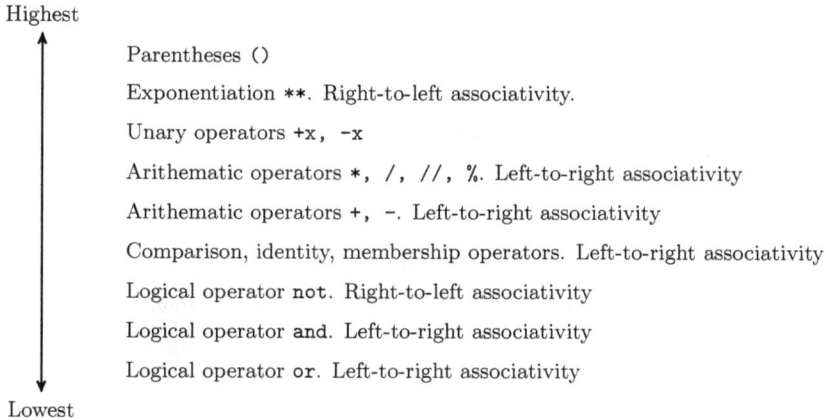

FIGURE 1.2: Precedence of operators.

The Python operators discussed earlier follow a specific precedence, which dictates the order in which operations are evaluated within an expression. Figure 1.2 illustrates the precedence levels of these operators. Parentheses have the highest precedence and can be used to explicitly group expressions, ensuring clarity and correctness. If you are uncertain about the precedence of an operator, using parentheses is a reliable way to control the evaluation order.

1.5 Control statements and loops

In computer programming, control statements are used to manage the flow of execution within a program, while loops are used to repeatedly execute a block of statements. In Python, control statements include if, if-else, and if-elif-else, which allow for conditional execution based on specific criteria. Loops in Python include the for loop, which iterates over a sequence, and the while loop, which repeats as long as a given condition remains true. Together, these constructs enable flexible and efficient program logic.

The following piece of code illustrates the usage of the control statements:

```
x = 1
if x > 0:
    print("x is positive")

y = 2
if x > y:
    print("x is greather than y")
```

```
 8 else :
 9     print ("x is not greater than y")
10
11 if y > 2:
12     print ("y is greater than 2")
13 elif y == 2:
14     print ("y is equal to 2")
15 else :
16     print ("y is less than 2")
```

Executing the above block of code in Spyder gives the following output:

```
1 x is positive
2 x is not greater than y
3 y is equal to 2
```

In the code above, the statements inside the if block are indented. A key feature of Python is its use of indentation to structure and organize code. In the example above, we used 4 spaces for indentation, which is the most common and widely recommended practice. However, you can choose to use a different number of spaces for indentation, as long as you maintain consistency throughout your code. Consistent indentation is crucial for ensuring readability and avoiding syntax errors in Python.

In the following code, we use a **for** loop to calculate the sum of integers 1 to 10:

```
1 dSum = 0
2 for i in range (1 , 11):
3     dSum += i
4 print (dSum)
```

When you execute the block of code above in Spyder, the result displayed will be 55. In this code, we utilized the built-in function **range**, which generates a sequence of integers starting from the specified start number (inclusive) up to the stop number (exclusive). This function is commonly used to control iterations in loops.

We can also use a **while** loop to calculate the sum of integers from 1 to 10 as follows:

```
1 dSum = 0
2 i = 0
3 while i < 10:
4     i += 1
5     dSum += i
6 print (dSum)
```

The output of the above code is also 55.

All the control statements and loops can be nested. For example, we use an `if` statement inside a `for` loop in the following code:

```
dSum = 0
for i in range(1, 1001):
    dSum += i
    if i % 100 == 0:
        print("Iteration {}: {}".format(i, dSum))
```

In the above code, we use a `for` loop to calculate the sum from 1 to 1000 and display the result every 100 steps. Executing the above code gives the following output:

```
Iteration 100: 5050
Iteration 200: 20100
Iteration 300: 45150
Iteration 400: 80200
Iteration 500: 125250
Iteration 600: 180300
Iteration 700: 245350
Iteration 800: 320400
Iteration 900: 405450
Iteration 1000: 500500
```

Python includes two key statements for controlling loops: **break** and **continue**. The **break** statement enables a loop to exit prematurely, stopping further iterations entirely. On the other hand, the **continue** statement skips the remaining code in the current iteration and proceeds directly to the next iteration of the loop. These statements provide greater flexibility in managing loop behavior and flow. The following piece of code illustrates the usage of the **break** statement:

```
dSum = 0
i = 0
while True:
    i += 1
    dSum += i
    if i >= 10:
        break
print(dSum)
```

In the code above, we implement an infinite loop, which continues indefinitely until the condition i >= 10 is met, at which point the loop terminates. The output of this code is 55. This example demonstrates how an infinite loop can be controlled using a conditional statement to ensure it exits at the appropriate time.

The following code shows how to use the **continue** statement to skip some statements in a loop:

```
1  dSum = 0
2  for i in range(1, 11):
3      if i % 2 == 1:
4          continue
5      dSum += i
6  print(dSum)
```

In the code above, when i is an odd number, the statement dSum += i is skipped during that iteration. The code sums all even numbers between 1 and 10, resulting in a final value of 30.

It's important to note that the **break** and **continue** statements only impact the innermost loop that encloses them. They do not affect any outer loops or other control structures in the program.

1.6   Functions

In Python, functions can be defined to encapsulate blocks of reusable code. By using functions, we can reduce code duplication, improve readability, and enhance the maintainability of our programs. Functions allow us to organize code into modular components, making it easier to debug, test, and reuse.

The following block of code illustrates the syntax of defining a function in Python:

```
1  def sqrt(x):
2      if x < 0:
3          return("{} is negative".format(x))
4      else:
5          return(math.sqrt(x))
6
7  print(sqrt(-1))
8  print(sqrt(2))
```

In the first line, **def** is a keyword used to define a function, **sqrt** is the name of the function, and x within the parentheses represents an argument passed to the function. This structure allows the function to accept input and perform operations based on that input. Executing the above block of code in Spyder gives the following output:

```
1  -1 is negative
2  1.4142135623730951
```

In the code above, we utilized the Python package **math**. To ensure the code works, we must first import the package by running the following line of code:

```
1 import math
```

This statement makes the functions and constants provided by the **math** package available for use in our program.

Python supports a special type of function called lambda functions, also known as anonymous functions. A lambda function is a concise, nameless function typically used for short, simple operations where a full function definition is unnecessary. Lambda functions are particularly useful for quick, inline functionality. The following code illustrates the definition and the use of a lambda function:

```
1 add = lambda x, y: x+y
2 print(add(1,2))
```

In the first line of the code above, **lambda** is the keyword used to define an anonymous function, the expression **x, y** represents the arguments separated by a comma, and the expression **x + y** represents the operation performed by the function. When this block of code is executed, the result is 3. Lambda functions provide a compact way to define simple functions inline.

The following block of code shows how to use a lambda function to create a list:

```
1 x = [(lambda x: math.sqrt(x))(i) for i in range(1,11)]
2 print(x)
```

The above code creates a list $[\sqrt{1}, \sqrt{2}, \ldots, \sqrt{10}]$. Executing the above block of code gives the following output:

```
1 [1.0, 1.4142135623730951, 1.7320508075688772, 2.0,
    2.23606797749979, 2.449489742783178,
    2.6457513110645907, 2.8284271247461903, 3.0,
    3.1622776601683795]
```

1.7 File IO

Reading and writing files are fundamental tasks in any programming language. Table 1.11 provides a list of commonly used functions for file handling in Python. When opening a file, it is important to ensure that Python can locate the file. This can be achieved by either placing the file in the working directory or specifying the full path to the file as an argument in the **open** function.

TABLE 1.11: Some functions for handling files.

Function	Description
open	Opens a file for reading or writing
close	Closes an opened file
read	Reads the content of an opened file
readline	Reads a single line from an opened file
readlines	Reads all lines from an opened file
write	Writes a string to an opened file
writelines	Writes a list of strings to an opened file

Suppose that we have a file named `file.txt` in the directory `C:\Users\`
`gjgan\Documents\ResearchU\book\dcpython\code`. The content of this file
is

```
Python is
a general purpose
programming language.
```

To read this file in Python, we can proceed as follows:

```
import os
fi = open(os.path.join(r"c:\users\gjgan\documents\researchu
    \book\dcpython\code", "file.txt"), "r")
content = fi.read()
fi.close()
print(content)
```

In the code above, we used the `os` package to combine the file path and the
file name. The symbol `r` before the path instructs Python to interpret the
string as a raw string. In a raw string, backslashes (\) are treated not as
escape sequences but as literal characters. Additionally, the argument ''r''
specifies that the file is opened in read mode, allowing the program to access
the file's contents for reading. Executing the above block of code gives the
following output:

```
Python is
a general purpose
programming language.
```

When reading from or writing to a file is complete, it is important to
ensure the file is properly closed by calling the `close` function. However, to
avoid manually closing the file, we can use the `with` statement. This approach
automatically closes the file once the block of code within the `with` statement
is executed, ensuring proper file handling and reducing the risk of resource
leaks. The following block of code illustrates the use of the `with` statement:

```
1  with open(os.path.join(r"c:\users\gjgan\documents\researchu
       \book\dcpython\code", "file.txt"), "r") as fi:
2      content = fi.read()
3  print(content)
```

The **read** function reads the entire content of a file into a single string. If we want to read the file into a list of lines, we can use the **readlines** function. This function is particularly useful when we need to process or manipulate the file content line by line. The following block of code illustrates the **readlines** function:

```
1  with open(os.path.join(r"c:\users\gjgan\documents\researchu
       \book\dcpython\code", "file.txt"), "r") as fi:
2      content = fi.readlines()
3  print(content)
```

Executing the above block of code gives the following output:

```
1  ['Python is\n', 'a general purpose\n', 'programming
      language.']
```

The symbol \n is the line break symbol.

Writing to a file can be performed in two distinct modes: overwriting the existing content or appending to it. To overwrite a file, open it with the mode argument ``w``. This will replace the file's current content with the new data. To append to a file, open it with the mode argument ``a``. This will add new content to the end of the file without altering the existing content. For example, the following block of code writes two lines to a file named **integers.txt**:

```
1  with open(os.path.join(r"c:\users\gjgan\documents\researchu
       \book\dcpython\code", "integers.txt"), "w") as fi:
2      fi.write(",".join([str(i) for i in range(1,6)]))
3      fi.write("\n")
4      fi.write(",".join([str(i) for i in range(6,11)]))
```

In the code above, the function **str** is a built-in function of Python that is used to convert an object to a string. After executing the above block of code, we can open the file **integers.txt** in a text editor and see the following content:

```
1  1,2,3,4,5
2  6,7,8,9,10
```

1.8   Error handling

Python offers a mechanism to handle errors or exceptions that may occur during program execution. This is accomplished using the `try-except` statement. The `try` block contains the code that might raise an exception, while the `except` block defines how to handle the exception if it occurs. This approach ensures that the program can gracefully manage unexpected issues without crashing.

The following block of code illustrates the use of the `try-except` statement:

```
1  x = -2
2  try:
3      print(math.sqrt(x))
4  except ValueError as e:
5      print(str(e))
```

In the code above, `ValueError` is a built-in exception of Python. Executing the above block of code in Spyder gives the following output:

```
1  math domain error
```

Python provides many built-in exceptions such as `ValueError`, `TypeError`, `RuntimeError`, `ZeroDivisionError`, `IndexError`, etc. To raise an error manually, we can use the keyword **raise**. The following block of code shows how to raise an error manually:

```
1  def sqrt2(x):
2      if not isinstance(x, (int, float)):
3          raise ValueError("{} is not numeric".format(x))
4      elif x < 0:
5          raise ValueError("{} is negative".format(x))
6      else:
7          return(math.sqrt(x))
8
9  for x in [-2, 2, "a"]:
10     try:
11         print(sqrt2(x))
12     except ValueError as e:
13         print(str(e))
```

In the code above, we define a function called `sqrt2`. The function begins by verifying whether the input is numeric and then checks if it is negative. If the input satisfies both conditions, the function calculates and returns the square root of the input. Executing the above block of code gives the following output:

```
1  -2 is negative
2  1.4142135623730951
3  a is not numeric
```

Handling exceptions ensures that the `for` loop completes all iterations. Without exception handling, the loop may terminate prematurely when an error occurs.

1.9 Object-oriented programming

Python supports object-oriented programming (OOP), which is a programming paradigm that uses objects to structure programs. An object models a real-world entity and can contain data and methods. Object-oriented programs are more maintainable and scalable than non-object-oriented programs.

In OOP, an object is instantiated from a class, which serves as a template defining the attributes and behaviors shared by that type of object. OOP is built on three core principles: encapsulation, inheritance, and polymorphism. Encapsulation integrates data and methods into a single unit, the class, restricting direct access to certain details. Inheritance allows a new class to derive from an existing class, inheriting its data and methods. Polymorphism enables a single interface to represent different underlying implementations, allowing for flexible and dynamic behavior.

The following piece of code illustrates how to define classes in Python:

```
1  from abc import abstractmethod
2  class Algorithm:
3      def __init__(self, name):
4          self.name = name
5      @abstractmethod
6      def fit(self):
7          pass
8
9  class Kmean(Algorithm):
10     def fit(self):
11         return("Clustering by kmeans")
12
13 class Kmode(Algorithm):
14     def fit(self):
15         return("Clustering by kmodes")
16
17 alg1 = Kmean("algorithm 1")
18 alg2 = Kmode("algorithm 2")
19
20 print(alg1.fit())
```

```
21  print(alg2.fit())
22  print(alg1.name)
23  print(alg2.name)
```

In the first line of the code above, we import the decorator @abstractmethod from the package abc, which defines abstract base classes in Python. This decorator is used to indicate that a method is an abstract method.

In Lines 2–7, we define a class with a property and an abstract method. In Lines 9–15, we define two subclasses of the class Algorithm. The abstract method in the parent class is implemented in the subclasses. In Lines 17–18, we create objects of the two classes. In Lines 20–23, the method is called and the property is displayed. Executing the above block of code gives the following output:

```
1  Clustering by kmeans
2  Clustering by kmodes
3  algorithm 1
4  algorithm 2
```

In Python, the __init__ method is a special method that serves as the constructor of a class. It is used to initialize objects of a class. This method is called immediately after an object is created.

1.10   Code Optimization

Unlike compiled languages such as C, C++, and Java, Python is a scripting language. Consequently, its performance on certain computational tasks may be slower compared to compiled languages. One way to enhance Python's performance is by using Cython, which allows Python code to be converted into compiled C code with C data types.

To illustrate how to use Cython to compile Python code to binary code, let us consider the calculation of the nth term of the Fibonacci sequence. The Fibonacci sequence $\{x_n\}_{n \geq 1}$ is defined recursively as follows:

$$x_1 = x_2 = 1, \quad x_n = x_{n-1} + x_{n-2}, \quad n \geq 3. \tag{1.1}$$

In Python, we can define a nested function to calculate the $n$term of the Fibonacci sequence:

```
1  def fibo(n):
2      if n == 1 or n == 2:
3          return(1)
4      else:
5          return(fibo(n-1) + fibo(n-2))
```

```
 6
 7 import time
 8 beg = time.time()
 9 print(fibo(40))
10 end = time.time()
11 print("Execution time: {:.4f} seconds".format(end - beg))
```

The nested function is defined in Lines 1–5 of the above block of code. The function fibo calls itself repeatedly to calculate the nth term by using the definition given in Equation (1.1). The code in Lines 7–11 is used to test the performance of the nested function. Executing the above code in a laptop computer produced the following output:

```
1 102334155
2 Execution time: 73.5849 seconds
```

From the output, we see that it took the nested function about 73.5849 seconds to calculate the 40th term.

To improve the performance, we can convert the nested function in Python to binary code. To do that, we need to create two Python files. The first file is named fibonacci.py and contains the nested function:

```
1 def fibo(n):
2     if n == 1 or n == 2:
3         return(1)
4     else:
5         return(fibo(n-1) + fibo(n-2))
```

The second file is named **setup.py** and contains the following code:

```
1 from setuptools import setup
2 from Cython.Build import cythonize
3
4 setup(
5     ext_modules=cythonize("fibonacci.py"),
6 )
```

Suppose that the two files are saved to the working directory. Then we open the Anaconda Prompt and change the directory to the working directory. To compile the Python code, we execute the following commands in the Anaconda Prompt:

```
1 conda activate dc
2 python setup.py build_ext --inplace
```

The first command is used to activate the Python environment dc, which contains the **Cython** package (see Section 1.1). The second command does the compilation. In Windows, the resulting compiled file is a .pyd file. In Linux, the resulting compiled file is a .so file.

Once the compiled file is created, we can use the nested function in Python
as follows:

```
1  import fibonacci
2  beg = time.time()
3  print(fibonacci.fibo(40))
4  end = time.time()
5  print("Execution time: {:.4f} seconds".format(end - beg))
```

Executing the above code in the same laptop produced the following output:

```
1  102334155
2  Execution time: 22.9853 seconds
```

From the output, we see that the compiled function was more than three times
faster than the pure Python function.

1.11   Summary

In this chapter, we introduced the fundamentals of Python programming. In
particular, we introduced data structures, operators, flow control statements,
user-defined functions, exception handling, and object-oriented programming.
For further readings, readers are referred to [12] and [172], which cover an
introduction to Python programming and its applications in machine learning.
For more information about object-oriented programming, readers are referred
to [235].

2

The NumPy Library

NumPy stands for Numerical Python and is a fundamental library of Python for scientific computing. It is extensively utilized in fields such as machine learning and data science. In this chapter, we provide an introduction to NumPy.

2.1 Arrays

An array is a data structure designed to store a collection of elements of the same data type. Elements within an array can be efficiently accessed and modified using their indices. NumPy offers a wide range of functions for creating and manipulating multidimensional arrays, making it a powerful tool for scientific computing.

To use NumPy, we first need to install it. This can be done at the same time when we create the Python environment (see Section 1.1). Once it is installed, we can import it by executing the following code:

```
import numpy as np
```

In the above code, the **as** keyword is used to give **numpy** a different name **np**. The short name makes the code more readable and concise.

In NumPy, the **array** function can be used to create arrays. For example, the following piece of code creates three vectors:

```
v1 = np.array([1, 2, 3])
v2 = np.array([[1, 2, 3]])
v3 = np.array([[1], [2], [3]])

print(v1.shape)
print(v2.shape)
print(v3.shape)
```

In the above code, v1 is a one-dimensional array, v2 is a row vector, and v3 is a column vector. The **shape** property tells the dimension of the array. Executing the above block of code in Spyder gives the following output:

```
1 (3,)
2 (1, 3)
3 (3, 1)
```

The row vector and the column vector are special matrices.

The following block of code illustrates how to create 2-dimensional arrays:

```
1 m1 = np.array([[1, 2], [3, 4]])
2 m2 = np.array(range(1, 5)).reshape((2,2))
3
4 print(m1)
5 print(m2)
```

The first 2-dimensional array is created by specifying all the rows in a list. The second 2-dimensional array is created by applying the **reshape** function to a list. Executing the above block of code in Spyder gives the following output:

```
1 [[1 2]
2 [3 4]]
3 [[1 2]
4 [3 4]]
```

From the output, we see that the **reshape** function divides the list of numbers into rows of numbers.

Creating high-dimensional arrays is similar as shown in the following block of code:

```
1 t1 = np.array([[[1, 2], [3, 4]], [[5, 6], [7, 8]]])
2 t2 = np.array(range(1, 9)).reshape((2, 2, 2))
3
4 print(t1)
5 print(t1.shape)
6 print(t2)
7 print(t2.shape)
```

The output of the above block of code is

```
1 [[[1 2]
2 [3 4]]
3
4 [[5 6]
5 [7 8]]]
6 (2, 2, 2)
7 [[[1 2]
8 [3 4]]
9
10 [[5 6]
11 [7 8]]]
12 (2, 2, 2)
```

The two arrays created above are 3-dimensional arrays. These high-dimensional arrays are called tensors.

Multidimensional arrays can be converted to one-dimensional arrays by using the `flatten` function. For example, the tensor `t1` can be converted to a one-dimensional array as follows:

```
print(t1.flatten())
```

The output is

```
[1 2 3 4 5 6 7 8]
```

NumPy also provides functions to create special arrays such as arrays with constant values. The following block of code illustrates the use of such functions:

```
s1 = np.zeros((2, 2))
s2 = np.ones((2, 2))
s3 = np.full((2, 2), 3)

print(s1)
print(s2)
print(s3)
```

The output of the above block of code is

```
[[0. 0.]
 [0. 0.]]
[[1. 1.]
 [1. 1.]]
[[3 3]
 [3 3]]
```

The `linspace` function can be used to create arithmetic sequences. The following block of code shows the usage of this function:

```
s4 = np.linspace(0, 1, 10)
s5 = np.linspace(1, 10, 10)
print(s4)
print(s5)
```

The output of the above block of code is

```
[0.         0.11111111 0.22222222 0.33333333 0.44444444
 0.55555556
0.66666667 0.77777778 0.88888889 1.         ]
[ 1.  2.  3.  4.  5.  6.  7.  8.  9. 10.]
```

2.2   Array indexing and slicing

Selecting elements of an array can be done by indexing and slicing in the same way as for a list. The methods given in Table 1.2 can be used for accessing elements in arrays.

The following block of code illustrates various methods for selecting elements from a one-dimensional array:

```
v1 = np.array(range(1, 9))
print(v1)
print(v1[:]) # select all elements
print(v1[:3]) # select elements with indices < 3
print(v1[3:]) # select elements with indices >= 3
print(v1[3]) # select the element with index 3
print(v1[[0,2,4]]) # select elments with indices in [0,2,4]
print(v1[-1]) # select the last element
print(v1[-3:-1]) # select elements with indices -3, -2
print(v1[-3:]) # select the last three elements
```

In the above code, how the meaning of each line is explained in the comment. Executing the above block of code gives the following output:

```
[1 2 3 4 5 6 7 8]
[1 2 3 4 5 6 7 8]
[1 2 3]
[4 5 6 7 8]
4
[1 3 5]
8
[6 7]
[6 7 8]
```

The indices of a one-dimensional array are similar to those of a list. Let v be a one-dimensional array with n elements. Then the indices of the elements can be $0, 1, \ldots, n - 1$ or $-n, -(n - 1), \ldots, -1$. As a result, elements in an array can be accessed by the method [start:stop:stride] (see Table 1.2). If stride is positive, the elements are selected from left to right. If stride is negative, the elements are selected from right to left. The value of stride cannot be zero. The following block of code shows the use of this method for selecting elements:

```
print(v1[::1]) # select elements with indices 0,1,2,...
print(v1[::2]) # select elements with indices 0, 2, 4, ...
print(v1[::-1]) # select elements with indices -1, -2, ...
print(v1[0::-1]) # select the element with index 0
print(v1[5::-1]) # select elements with indices 5,4,...,0
print(v1[-3::-1]) # select elements with indices -3,-4,...
```

The output of the above block of code is

```
1  [1 2 3 4 5 6 7 8]
2  [1 3 5 7]
3  [8 7 6 5 4 3 2 1]
4  [1]
5  [6 5 4 3 2 1]
6  [6 5 4 3 2 1]
```

Selecting elements from a two-dimensional array requires two coordinates. The following block of code shows how to select elements from a two-dimensional array:

```
1  m1 = np.array(range(1,9)).reshape((2,4))
2
3  print(m1)
4  print(m1[:,1]) # select the second column
5  print(m1[0,:]) # select the first row
6  print(m1[1,::-1]) # select the second row reversed
7  print(m1[0,[1,3]]) # select the first row and the second,
      fourth columns
```

The output of the above block of code is

```
1  [[1 2 3 4]
2   [5 6 7 8]]
3  [2 6]
4  [1 2 3 4]
5  [8 7 6 5]
6  [2 4]
```

2.3   Views and copies

In NumPy, the array data structure consists of two parts: the contiguous data buffer that stores the actual data elements and the metadata that contains information (e.g., data type, shape) about the data buffer. To improve performance, NumPy often uses views rather than copies when operating on arrays. A view of an array is a way of looking at the data by changing certain metadata of the array without changing the actual data buffer. A copy of an array is a duplication of the data buffer and the metadata. Changes made to a view of an array will reflect in the original array. However, changes made to a copy of an array will not affect the original array.

NumPy has certain rules about whether to create a view or a copy. When elements in a selection of an array can be addressed by offsets and strides in the original array, a view will be created. The following block of code shows that views are created:

```
m = np.array(range(1,7)).reshape((3,2))

print(m)
v1 = m[0:2,] # this is a view
v1[0, 1] = -1
print(np.may_share_memory(v1, m))
print(m)

v2 = m[0:3:2,] # this is a view
v2[0, 1] = -2
print(np.may_share_memory(v2, m))
print(m)
```

In the above code, we select subsets from an array and make changes to the subsets. We use the NumPy function **may_share_memory** to check whether the subsets share memory with the original array. Executing the above block of code in Spyder gives the following output:

```
[[1  2]
 [3  4]
 [5  6]]
True
[[ 1  -1]
 [ 3   4]
 [ 5   6]]
True
[[ 1  -2]
 [ 3   4]
 [ 5   6]]
```

From the output, we see that views were created and changes made to the views were reflected in the original array.

In NumPy, advanced indexing always creates copies. The following block of code illustrates this:

```
m = np.array(range(1,7)).reshape((3,2))

print(m)
c1 = m[[0,2],:] # this is a copy
c1[0, 1] = -1
print(np.may_share_memory(v1, m))
print(m)
```

Executing the above block of code gives the following output:

```
[[1 2]
 [3 4]
 [5 6]]
False
[[1 2]
 [3 4]
 [5 6]]
```

From the output, we see that the change made to the copy did not change the original array.

Additionally, we can use the **base** attribute to check whether an array is a view or a copy of another array. The following block of code illustrates this:

```
m = np.array(range(1,7)).reshape((3,2))
v1 = m[0:2,]
c1 = m[[0,1],]
print(v1.base)
print(c1.base)
```

The output of the above block of code is

```
[1 2 3 4 5 6]
None
```

If the **base** attribute is **None**, then it is a copy.

2.4   Array operations

We can perform numerical operations on NumPy arrays. For example, we can apply the arithmetic operators described in Section 1.4 on NumPy arrays. All the operations are element-wise. The following block of code illustrates these operations on NumPy arrays:

```
m1 = np.array(range(1,5)).reshape((2,2))
m2 = np.array(range(5,9)).reshape((2,2))

print(m1 + m2)
print(m1 - m2)
print(m1 * m2)
print(m1 / m2)
print(m1 ** m2)
print(m1 + 2)
print(m1 ** 2)
```

Executing the above block of code in Spyder gives the following output:

```
[[ 6   8]
 [10  12]]
[[-4  -4]
 [-4  -4]]
[[ 5  12]
 [21  32]]
[[0.2          0.33333333]
 [0.42857143 0.5         ]]
[[    1      64]
 [ 2187  65536]]
[[3 4]
 [5 6]]
[[ 1   4]
 [ 9  16]]
```

If two arrays have the same shape, then it is straightforward to perform arithmetic operations on them. If two arrays have different shapes, operations can still be applied to them if the large array is a multiply of the short array. This is the broadcasting feature of NumPy. The following block of code illustrates this feature:

```
m1 = np.array(range(1, 9)).reshape((2,4))
v1 = np.array(range(1, 5))

print(m1)
print(v1)
print(m1 + v1)
```

The array `v1` and the array `m1` have different shapes. However, the small array will be repeated to match the large array. The output of the above block of code is

```
[[1 2 3 4]
 [5 6 7 8]]
[1 2 3 4]
[[ 2   4   6   8]
 [ 6   8  10  12]]
```

2.5 Functions

NumPy provides a rich set of functions that can be used on arrays. These functions include mathematical functions, statistical functions, and functions used for sorting, searching, and counting. In this section, we introduce a few functions provided by NumPy. For a complete list of functions provided by NumPy, readers are referred to the NumPy API reference at `https://numpy.org/`.

NumPy provides functions for generating random numbers, which are important for statistical modeling and simulation. The following block of code shows how to generate random numbers from different distributions:

```
rng = np.random.default_rng()

v1 = rng.random(10) # generate 10 random numbers from the
    uniform distribution on [0,1)
v2 = rng.standard_normal(10) # generate 10 random numbers
    from the standard normal distribution
v3 = rng.integers(0, 100, 10) # generate 10 random integers
    from {0,1,...,99}

print(v1)
print(v2)
print(v3)
```

The three random number generation functions are commonly used in practice. The first function `random` generates uniformly distributed random numbers. The second function `standard_normal` generates random numbers from the standard normal distribution, which has a mean of zero and a standard deviation of one. The third function generates random integers from an interval $[L, H)$. The low end of the interval is inclusive and the upper end of the interval is exclusive. The output of executing the above block of code looks like:

```
[0.90658443 0.99689331 0.22248615 0.06889595 0.09038821
    0.99628482
0.66239361 0.87383429 0.83905642 0.14733118]
[-0.7599299  -1.09448303  1.36601884 -1.98679173
    -0.19144456 -0.56262858
-0.64179878 -0.88431171  0.24020192 -1.29943219]
[88 67 55 81 36 98 29 85 97 82]
```

Note that you will see different outputs when you run the above block of code again. To repeat the random numbers, we need to fix a seed for the random number generator.

To fix the seed for the default random number generator, we can call the function as follows:

```
rng = np.random.default_rng(seed=2024)
print(rng.random(5))
rng = np.random.default_rng(seed=2024)
print(rng.random(5))
```

In the above code, we set the seed of the random number generator to be the same in two calls. If we execute the above block of code in Spyder, we see the following output:

```
[0.67583134 0.2143232   0.30945203 0.7994661   0.9958021 ]
[0.67583134 0.2143232   0.30945203 0.7994661   0.9958021 ]
```

From the output, we see that the outputs from the two calls are the same. You should be able to repeat the above outputs by executing the same block of code given above.

Some commonly used statistical functions are provided by NumPy. The following block of code illustrates the use of these statistical functions:

```
rng = np.random.default_rng(seed=1)
y = rng.standard_normal(1000)
print("97.5% percentile:", np.percentile(y, 97.5))
print("0.975 quantile:", np.quantile(y, 0.975))
print("min:", np.min(y))
print("median:", np.median(y))
print("max:", np.max(y))
print("mean:", np.mean(y))
print("std:", np.std(y))
```

In the above block of code, we generate an array of 1000 random numbers from the standard normal distribution and calculate summary statistics. The output of the code is given below:

```
97.5% percentile: 1.8609122904322217
0.975 quantile: 1.8609122904322217
min: -3.5488049709979372
median: -0.032839126821055185
max: 3.7516349672663583
mean: -0.05425322276336561
std: 0.9862611378496257
```

From the output, we see that the two functions **percentile** and **quantile** produce the same result if we supply the correct argument values.

NumPy also provides functions for sorting, searching, and counting. For example, the following block of code shows how to sort and search elements of arrays:

```
rng = np.random.default_rng(seed=1)
y = rng.standard_normal(5)

print(y)
print(np.sort(y))
print(np.argsort(y))
```

The **sort** function sorts an array in an increasing order and returns the sorted array. The **argsort** function returns an array of indices instead of the values. The output of the above block of code is

```
1 [ 0.34558419   0.82161814   0.33043708  -1.30315723
     0.90535587]
2 [-1.30315723   0.33043708   0.34558419   0.82161814
     0.90535587]
3 [3 2 0 1 4]
```

If we just need to know the maximum value, the minimum value, and their indices, we can use the functions **max**, **min**, **argmax**, and **argmin** as shown in the following block of code:

```
1 rng = np.random.default_rng(seed=2)
2 y = rng.standard_normal(5)
3
4 print(y)
5 print(np.max(y))
6 print(np.argmax(y))
7 print(np.min(y))
8 print(np.argmin(y))
```

The output of the code is

```
1 [ 0.18905338  -0.52274844  -0.41306354  -2.44146738
     1.79970738]
2 1.799707382720902
3 4
4 -2.4414673826398556
5 3
```

2.6   Matrices

Matrices are special two-dimensional arrays. NumPy provides many functions for matrix calculations. Creating a matrix can be done by using the **matrix** function on a two-dimensional array. The following block of code illustrates the difference of NumPy arrays and matrices:

```
1 m1 = np.array([1, 0.1, 0.1, 1]).reshape((2,2))
2 m2 = np.matrix(m1)
3
4 print(type(m1))
5 print(type(m2))
6
```

```
 7 print(m1 * m1) # element-wise multiplication
 8 print(m2 * m2) # matrix multiplication
 9 print(m1 @ m1) # matrix multiplication
10 print(m2 @ m2) # matrix multiplication
```

Executing the above block of code in Spyder gives the following output:

```
 1 <class 'numpy.ndarray'>
 2 <class 'numpy.matrix'>
 3 [[1.    0.01]
 4  [0.01 1.  ]]
 5 [[1.01 0.2 ]
 6  [0.2  1.01]]
 7 [[1.01 0.2 ]
 8  [0.2  1.01]]
 9 [[1.01 0.2 ]
10  [0.2  1.01]]
```

From the output, we see that the multiplication operator * is interpreted differently for arrays and matrices. However, the operator @ is interpreted as matrix multiplication for both arrays and matrices.

The following block of code shows some examples of the matrix functions:

```
 1 v = np.array(range(1,5))
 2 m = v.reshape((2,2))
 3
 4 print("diag:", np.diag(m))
 5 print("lower triangle:\n", np.tril(m))
 6 print("upper triangle:\n", np.triu(m))
 7 print("transpose:\n", np.transpose(m))
 8 print("rank:", np.linalg.matrix_rank(m))
 9 print("norm of v:", np.linalg.norm(v))
10 print("norm of m:", np.linalg.norm(m))
11 print("trace:", np.linalg.trace(m))
```

Executing the block of code gives the following output:

```
 1 diag: [1 4]
 2 lower triangle:
 3  [[1 0]
 4  [3 4]]
 5 upper triangle:
 6  [[1 2]
 7  [0 4]]
 8 transpose:
 9  [[1 3]
10  [2 4]]
11 rank: 2
12 norm of v: 5.477225575051661
```

```
13 norm of m: 5.477225575051661
14 trace: 5
```

Most of the functions are straightforward to understand. The **norm** function can calculate different norms. The default norm is the Frobenius norm, which is defined to be the square root of the sum of the squared elements of an array.

NumPy has functions for solving linear equation systems. Consider the following linear equation system

$$x - 2y + 3z = 7,$$
$$2x + y + z = 4,$$
$$-3x + 2y - 2z = -10.$$

This linear equation system can be solved by NumPy functions as follows:

```
1 A = np.matrix(np.array([1, -2, 3, 2, 1, 1, -3, 2, -3]).
    reshape((3,3)))
2 b = np.matrix(np.array([7, 4, -10]).reshape((3,1)))
3 x = np.linalg.solve(A, b)
4
5 print(x)
6 print(A * x)
```

Executing the above block of code gives the following output:

```
1 [[ 1.5]
2 [-0.5]
3 [ 1.5]]
4 [[  7.]
5 [  4.]
6 [-10.]]
```

The above simple linear equation system can also be solved by multiplying the inverse of the coefficient matrix and the constant vector. The following block of code shows this approach:

```
1 invA = np.linalg.inv(A)
2 print(invA * b)
```

Executing the above block of code gives the solution:

```
1 [[ 1.5]
2 [-0.5]
3 [ 1.5]]
```

2.7 File IO

NumPy offers many functions for saving arrays to files and loading arrays from files. In particular, the files can be text files, binary files, or compressed files.

The following block of code shows how to save an array to a text file and load the array from the text file:

```
rng = np.random.default_rng(seed=1)
dat = np.array(rng.standard_normal(10000)).reshape
    ((1000,10))

np.savetxt("dat.csv", dat, fmt="%.8f", delimiter=",")
dat2 = np.loadtxt("dat.csv", dtype=float, delimiter=",")
print(np.linalg.norm(dat2-dat))
```

In the above code, we create a random two-dimensional array, which contains 1000 rows and 10 columns. The **savetxt** function is called to save the array to a text file. All numbers are rounded to eight decimal places. Then the **loadtxt** function is called to load the array from the file. In the last line, the norm of the difference between the original array and the array loaded from the file is calculated. Executing the block of code gives the following output:

```
2.863391613525999e-07
```

The result shows that the array recovered from the file is almost the same as the original array.

Saving a large array to a text file may not be desirable as the file size can be large. NumPy provides functions to save arrays to binary files. The following block of code shows how to save the above array to a binary file and load the array from the file:

```
np.save("dat.npy", dat)
dat3 = np.load("dat.npy")
print(np.linalg.norm(dat3-dat))
```

The binary file has the **.npy** extension. Executing the above block of code gives the following output:

```
0.0
```

The result shows that the recovered array is exactly the same as the original array. In addition, the size of the binary file is smaller than that of the text file.

2.8 Code optimization

NumPy arrays are faster than Python lists for the following reasons. First, NumPy arrays are homogeneous data structures. A NumPy array contains data with the same data type. Second, the data contained in a NumPy array are stored in contiguous memory blocks. Third, most NumPy operations are written in the C language and are compiled. Since Python is a scripting language, there are still strategies to improve the performance of NumPy code.

The first strategy is to use vectorization over loops. To illustrate this, let us consider calculating the following sum:

$$\sum_{i=1}^{1000000} \frac{1}{\sqrt{i}}.$$

We can use a loop or the `sum` function provided by NumPy to calculate the sum. The following block of code shows the performance of the two approaches:

```
import timeit

mysetup = "from math import sqrt"
mycode = '''
dSum = 0
for i in range(1,1000001):
dSum += 1/sqrt(i)
'''
print(timeit.timeit(setup=mysetup, stmt=mycode, number=100)
    )

mysetup2 = "import numpy as np"
mycode2 = '''
dSum = np.sum(np.reciprocal(np.sqrt(np.r_[1:1000001])))
'''
print(timeit.timeit(setup=mysetup2, stmt=mycode2, number
    =100))
```

In the above code, we use the Python package `timeit` to measure the execution time of the two approaches. Executing the above block of code in Spyder, we see the following output:

```
29.838358900044113
2.259695300133899
```

From the results, we see that it took the loop about 29.84 seconds to calculate the sum 100 times. For the second approach that uses vectorization, it only took about 2.26 seconds to calculate the sum 100 times. The vectorization approach is much faster than the loop.

The second strategy is to use the broadcasting feature of NumPy. The broadcasting feature allows numerical operations to be performed on arrays with different shapes. NumPy's broadcasting rules are based on comparing the shapes of two arrays. Let $(a_1, a_2, \ldots, a_m)$ and $(b_1, b_2, \ldots, b_n)$ be the shapes of two arrays. When the two arrays have the same number of dimensions (i.e., $n = m$), the two arrays are broadcastable if $b_j = a_j$ or $\min\{a_j, b_j\} = 1$ for all $j = 1, 2, \ldots, n$. In other words, two dimensions are compatible if they are equal or one of them is one. When the two arrays have different dimensions (i.e., $n \neq m$), the missing dimensions of the small array will be assumed to be one in the following way. Suppose that $m < n$. Then the shape $(a_1, a_2, \ldots, a_m)$ will be extended to

$$(\underbrace{1, \ldots, 1}_{n-m}, a_1, a_2, \ldots, a_m),$$

which has the size of n. Then the aforementioned rule will be applied. That is, the two arrays are broadcastable if $b_j = a_{j-n+m}$ or $\min\{b_j, a_{j-n+m}\} = 1$ for all $j = n, n-1, \ldots, n-m+1$.

NumPy's broadcasting rules are different from those of R. For example, we cannot add the two arrays $[1, 2, 3, 4]$ and $[1, 2]$ in NumPy. However, we can add them in R.

The following block of code illustrates NumPy's broadcasting feature:

```
a1 = np.array(range(1,9)).reshape((2,2,2))
a2 = np.array(range(1,5)).reshape((2,2))

print(a1)
print(a2)
print(a1+a2)
```

The output of the above code is

```
[[[1 2]
 [3 4]]

 [[5 6]
 [7 8]]]
[[1 2]
 [3 4]]
[[[ 2  4]
 [ 6  8]]

 [[ 6  8]
 [10 12]]]
```

2.9 Summary

NumPy is the fundamental package of Python for scientific computing and is widely used in the fields of machine learning and data science. In this chapter, we presented an introduction to the NumPy package, including array creation, array indexing and slicing, and array operations. For more information about NumPy, readers are referred to the NumPy's API reference and [172].

3

The Pandas Library

The Pandas library is a popular Python library built on NumPy and is designed to work with tabular data. In this chapter, we introduce the data structures of Pandas and some functions for manipulating data.

3.1 Pandas series

Pandas provides two basic data structures: series and data frames. A series is a one-dimensional labeled array that can hold data of any type. A data frame is a two-dimensional array that looks like a table with rows and columns. In this section, we introduce the first data structure.

To create a series, we use the **Series** function provided in Pandas. The following piece of code shows how to create a series:

```
import pandas as pd
import numpy as np

s1 = pd.Series([3.14, "str", True, np.array([1,2]), np.nan
    ])
print(s1)
```

The series created in the above code contains five elements that have different data types. Executing the above piece of code in Spyder gives the following output:

```
0        3.14
1         str
2        True
3      [1, 2]
4         NaN
dtype: object
```

The default indices of the elements in a series start from 0.

We can supply indices when creating a series. For example, we can provide the index to a series as follows:

DOI: 10.1201/9781003592648-3

```
s2 = pd.Series([3.14, "str", True, np.array([1,2]), np.nan
    ], index=["a", "b", "c", "d", "e"])
print(s2)
```

Executing the above code gives the following output:

```
a        3.14
b         str
c        True
d       [1, 2]
e         NaN
dtype: object
```

A series can be created from a dictionary. The following piece of code shows this method:

```
s3 = pd.Series({0: 3.14, 1: "str", 2: True, 3: np.array
    ([1,2]), 4: np.nan})
print(s3)
```

The output of the above code is

```
0        3.14
1         str
2        True
3       [1, 2]
4         NaN
dtype: object
```

The values and the indices of a series can be accessed by the attributes **values** and **index**, respectively. The following code shows the use of the two attributes:

```
s1.values
s1.index
```

Executing the above code line by line in Spyder gives the following output:

```
In [665]: s1.values
Out[665]: array([3.14, 'str', True, array([1, 2]), nan],
    dtype=object)

In [666]: s1.index
Out[666]: RangeIndex(start=0, stop=5, step=1)
```

Selecting elements of a series can be done by using the index and the slicing methods used for lists (see Table 1.2). If the indices of a series are continuous integers starting from 0, then we can use the slicing method for a series in the same way as for a list. The series **s1** created above has indices that are continuous integers starting from 0. The following code shows how to select elements from **s1**:

```
print(s1[0]) # get the first element
print(s1[:3]) # get the first three elements
print(s1[0:4:2]) # get elements at indices 0, 2
print(s1[-1:-4:-2]) # get elements at positions -1, -3
```

Executing the above block of code gives the following output:

```
3.14
0      3.14
1       str
2      True
dtype: object
0      3.14
2      True
dtype: object
4       NaN
2      True
dtype: object
```

If a series does not use continuous integers as indices, then we need to use the slicing method along with the indices to select elements from the series. The following block of code shows how to select elements from **s2**:

```
print(s2[s2.index[0]]) # get the first element
print(s2[s2.index[:3]]) # get the first three elements
print(s2[s2.index[0:4:2]]) # get elements at index[0],
    index[2]
print(s2[-1:-4:-2]) # get elements at positions -1, -3
print(s2[s2.index[-1:-4:-2]]) # get elements at index[0],
    index[2]
```

Executing the above block of code in Spyder gives the following output:

```
3.14
a      3.14
b       str
c      True
dtype: object
a      3.14
c      True
dtype: object
e       NaN
c      True
dtype: object
e       NaN
c      True
dtype: object
```

It is interesting to see that negative integers are treated as positions.

To use integer positions to select elements from a series, we can use the `iloc` attribute of the series. The following block of code shows how to use the `iloc` attribute to select elements:

```
print(s2.iloc[0]) # get the first element
print(s2.iloc[:3]) # get the first three elements
print(s2.iloc[0:4:2]) # get elements at indices 0, 2
print(s2.iloc[-1:-4:-2]) # get elements at indices -1, -3
```

Executing the above block of code gives the following output:

```
3.14
a      3.14
b       str
c      True
dtype: object
a      3.14
c      True
dtype: object
e       NaN
c      True
dtype: object
```

The `iloc` attribute used above uses implicit indexing. The `loc` attribute uses explicit indexing. The following code illustrates the use of the `loc` attribute to select elements:

```
print(s2.loc["d"]) # get the element with index d
print(s2.loc[["a", "b"]]) # get the elements with indices a
    , b
```

Executing the above block of code gives the following output:

```
[1 2]
a      3.14
b       str
dtype: object
```

3.2 Pandas data frames

In Pandas, a data frame is a two-dimensional data structure where both rows and columns are indexed and labeled. The Pandas data frame is more general than the data frame in the R language. In R, a column of a data frame contains the same type of data. In Pandas, a column of a data frame can have different types of data.

A data frame can be created in various ways. It can be created from a dictionary of collections (e.g., series, NumPy one-dimensional arrays, lists) and a NumPy two-dimensional array. The following piece of code shows how to create a data frame from a dictionary of series:

```
dic = {
        "V1": pd.Series([1, 2, 3], index=["r1", "r2", "r3"])
        ,
        "V2": pd.Series(["str", [0, 1], np.nan], index=["r1"
            , "r2", "r3"])
    }
df = pd.DataFrame(dic)
print(df)
```

Executing the above block of code gives the following output:

```
    V1      V2
r1  1      str
r2  2    [0, 1]
r3  3      NaN
```

The row indices of the resulting data frame are the union of the indices of the series in the input dictionary. In the above example, the series in the dictionary have the same indices. The following piece of code gives an example when the series have different indices:

```
dic2 = {
        "V1": pd.Series([1, 2, 3], index=["r1", "r2", "r3"])
        ,
        "V2": pd.Series(["str", [0, 1], np.nan], index=["r4"
            , "r5", "r6"])
    }
df2 = pd.DataFrame(dic2)
print(df2)
```

Executing the above block of code gives the following output:

```
    V1      V2
r1  1.0     NaN
r2  2.0     NaN
r3  3.0     NaN
r4  NaN     str
r5  NaN   [0, 1]
r6  NaN     NaN
```

From the output, we see that missing values are used to fill the data frame when a series does not have certain indices.

The following block of code shows how to create a data frame from a NumPy two-dimensional array:

```
dat = np.array(range(1,9)).reshape((4,2))
rownames = ["r1", "r2", "r3", "r4"]
colnames = ["V1", "V2"]
df3 = pd.DataFrame(dat, index=rownames, columns=colnames)
print(df3)
```

The output of the above block of code is

```
     V1   V2
r1    1    2
r2    3    4
r3    5    6
r4    7    8
```

In the above example, row labels are set by using the **index** argument and column labels are set by using the **columns** argument. The column labels and the row labels can be accessed by the attributes **columns** and **index**, respectively. The following code shows how to get these labels of **df3**:

```
print(df3.columns)
print(df3.index)
```

Executing the above code gives the following output:

```
Index(['V1', 'V2'], dtype='object')
Index(['r1', 'r2', 'r3', 'r4'], dtype='object')
```

Accessing elements of a data frame can be done by using explicit indices or implicit indices. The following block of code illustrates how to select elements from a data frame:

```
print(df3["V1"]) # get column V1
print(df3[["V1", "V1"]]) # get column V1 twice
print(df3.loc[["r1", "r3"], "V2"]) # get specified rows and
      columns by explicint indices
print(df3.iloc[[0,2], 1]) # get specified rows and columns
      by implicit indices
```

Executing the above block of code in Spyder gives the following output:

```
r1    1
r2    3
r3    5
r4    7
Name: V1, dtype: int64
```

```
 6        V1   V1
 7 r1     1    1
 8 r2     3    3
 9 r3     5    5
10 r4     7    7
11 r1     2
12 r3     6
13 Name:  V2,  dtype:  int64
14 r1     2
15 r3     6
16 Name:  V2,  dtype:  int64
```

Elements can also be selected based on some criteria. For example, the following code shows how to select elements by criterion:

```
1 print(df3[df3["V1"] > 3])
2 print(df3[df3.V1 > 3])
```

The output of executing the above code is

```
1        V1   V2
2 r3     5    6
3 r4     7    8
4        V1   V2
5 r3     5    6
6 r4     7    8
```

3.3   Views and copies

Since Pandas data structures depend on NumPy arrays, operations on Pandas series and data frames may return views instead of copies of the original data (see Section 2.3). However, NumPy is more consistent and predictable than Pandas in creating views. In NumPy, changes in views are reflected in the original array. In Pandas, changes in views might not always reflect in the original data.

To address the unpredictable behavior of creating views, Pandas has provided the copy-on-write mechanism since version 1.5. The following block of code illustrates the case when this feature is turned off:

```
1 pd.set_option("mode.copy_on_write", False)
2
3 m = pd.DataFrame({"a": range(1, 4), "b": range(4, 7)})
4
5 print(m)
```

```
6  v1 = m.iloc[0:2, ]
7  print(v1._is_view)
8  print(v1._is_copy)
9  v1.iloc[0,1] = -1
10 print(m)
```

Executing the above block of code gives the following output:

```
1      a   b
2  0   1   4
3  1   2   5
4  2   3   6
5  True
6  <weakref at 0x000002936B843C40; to 'DataFrame' at 0
       x00000293687D44D0>
7      a   b
8  0   1  -1
9  1   2   5
10 2   3   6
11 C:\Users\gjgan\AppData\Local\Temp\ipykernel_21832
       \3244611799.py:9: SettingWithCopyWarning:
12 A value is trying to be set on a copy of a slice from a
       DataFrame
13
14 See the caveats in the documentation: https://pandas.pydata
       .org/pandas-docs/stable/user_guide/indexing.html#
       returning-a-view-versus-a-copy
15   v1.iloc[0,1] = -1
```

From the output, we see that v1 is a view. Changing the view changed the original data m. However, Pandas raised the warning `SettingWithCopyWarning`.

To fix the warning, we can set the copy-on-write feature to be true. The following block of code illustrates this case:

```
1  pd.set_option("mode.copy_on_write", True)
2
3  m = pd.DataFrame({"a": range(1, 4), "b": range(4, 7)})
4
5  print(m)
6  v1 = m.iloc[0:2, ]
7  print(v1._is_view)
8  print(v1._is_copy)
9  v1.iloc[0,1] = -1
10 print(v1._is_view)
11 print(v1._is_copy)
12 print(m)
```

Executing the above block of code gives the following output:

```
    a   b
0   1   4
1   2   5
2   3   6
True
<weakref  at  0x000002936B843290;  to  'DataFrame'  at  0
    x00000293687D6150>
False
<weakref  at  0x000002936B843290;  to  'DataFrame'  at  0
    x00000293687D6150>
    a   b
0   1   4
1   2   5
2   3   6
```

From the output, we see that `v1` was a view of the original data. After a change was made to `v1`, `v1` was converted into a copy. The original data was not affected. When the copy-on-write feature was set to be true, the warning `SettingWithCopyWarning` was not raised.

3.4   Data manipulation

Pandas provides many functions for manipulating data frames. In this section, we introduce some commonly used function for manipulating data.

Before introduce data manipulation functions, let us first load a dataset from the UCI machine learning repository by using the Python package `ucimlrepo`. This package is installed in the conda environment in Section 1.1. The following piece of code is used to load the auto MPG dataset [211]:

```
from ucimlrepo import fetch_ucirepo

auto_mpg = fetch_ucirepo(id=9)

X = auto_mpg.data.features
y = auto_mpg.data.targets

print(type(X))
print(X.columns)
print(X.iloc[:,0:4])
print(y)
```

Executing the above block of code loads the auto MPG dataset and gives the following output:

```
 1 <class 'pandas.core.frame.DataFrame'>
 2 Index(['displacement', 'cylinders', 'horsepower', 'weight',
        'acceleration', 'model_year', 'origin'], dtype='object
        ')
 3       displacement   cylinders   horsepower   weight
 4 0            307.0          8        130.0     3504
 5 1            350.0          8        165.0     3693
 6 2            318.0          8        150.0     3436
 7 3            304.0          8        150.0     3433
 8 4            302.0          8        140.0     3449
 9 ..             ...        ...          ...      ...
10 393          140.0          4         86.0     2790
11 394           97.0          4         52.0     2130
12 395          135.0          4         84.0     2295
13 396          120.0          4         79.0     2625
14 397          119.0          4         82.0     2720
15
16 [398 rows x 4 columns]
17         mpg
18 0      18.0
19 1      15.0
20 2      18.0
21 3      16.0
22 4      17.0
23 ..      ...
24 393    27.0
25 394    44.0
26 395    32.0
27 396    28.0
28 397    31.0
29
30 [398 rows x 1 columns]
```

The auto MPG dataset has seven features and contains 398 records. The features are saved in the Pandas data frame X and the target is saved in the Pandas data frame y.

To display some summary statistics of this dataset and check whether it contains missing values, we can use the following code:

```
1 print(X.describe())
2 print(X.isna().sum())
```

Executing the above code gives the following output:

```
1          displacement    cylinders    ...    model_year
                 origin
2 count      398.000000   398.000000    ...    398.000000
         398.000000
3 mean       193.425879     5.454774    ...     76.010050
           1.572864
```

```
 4 std        104.269838     1.701004    ...      3.697627
        0.802055
 5 min         68.000000     3.000000    ...     70.000000
        1.000000
 6 25%        104.250000     4.000000    ...     73.000000
        1.000000
 7 50%        148.500000     4.000000    ...     76.000000
        1.000000
 8 75%        262.000000     8.000000    ...     79.000000
        2.000000
 9 max        455.000000     8.000000    ...     82.000000
        3.000000
10
11 [8 rows x 7 columns]
12 displacement    0
13 cylinders       0
14 horsepower      6
15 weight          0
16 acceleration    0
17 model_year      0
18 origin          0
19 dtype: int64
```

From the output, we see that the feature `horsepower` contains 6 missing values.

Pandas provides multiple ways to add or delete a column from a data frame. The following block of code illustrates different ways to add or delete a column:

```
 1 X1 = X.copy()
 2 print(X1.columns)
 3 X1["mpg"] = y # add a column at the end
 4 print(X1.columns)
 5 X1.pop("mpg") # drop a column
 6 print(X1.columns)
 7 X1.insert(0, "mpg", y) # insert a column before the first
       column
 8 print(X1.columns)
 9 del X1["mpg"] # delete a column
10 print(X1.columns)
```

In the above code, we make a copy of the data frame before performing various operations. Executing the above block of code gives the following output:

```
 1 Index(['displacement', 'cylinders', 'horsepower', 'weight',
       'acceleration',
 2        'model_year', 'origin'],
 3       dtype='object')
 4 Index(['displacement', 'cylinders', 'horsepower', 'weight',
```

```
    'acceleration',
5       'model_year', 'origin', 'mpg'],
6     dtype='object')
7 Index(['displacement', 'cylinders', 'horsepower', 'weight',
    'acceleration',
8       'model_year', 'origin'],
9     dtype='object')
10 Index(['mpg', 'displacement', 'cylinders', 'horsepower', '
    weight',
11      'acceleration', 'model_year', 'origin'],
12    dtype='object')
13 Index(['displacement', 'cylinders', 'horsepower', 'weight',
    'acceleration',
14      'model_year', 'origin'],
15    dtype='object')
```

The above output shows that columns changed after adding or deleting a column.

To create a new column from existing ones, we can use the **assign** function, which is illustrated in the following code:

```
1 X1 = X.assign(displacement2=np.sqrt(X1["displacement"]))
2 print(X1.head())
```

The **assign** function always returns a copy of the data. Executing the above block of code gives the following output:

```
1    displacement  cylinders  horsepower  ...  model_year
       origin  displacement2
2 0        307.0          8       130.0  ...          70
         1    17.521415
3 1        350.0          8       165.0  ...          70
         1    18.708287
4 2        318.0          8       150.0  ...          70
         1    17.832555
5 3        304.0          8       150.0  ...          70
         1    17.435596
6 4        302.0          8       140.0  ...          70
         1    17.378147
7
8 [5 rows x 8 columns]
```

We can also add rows to and drop rows from a data frame. The following block of code illustrates how to add and drop rows:

```
1 n, d = X.shape
2 r = pd.DataFrame(np.array([X.iloc[np.random.randint(0, n),
    j] for j in range(d)]).reshape((1,d)), columns=X.
    columns)
```

```
 3
 4  X1 = X._append(r)  # add a row
 5  X1.loc[len(X1)] = r.values[0]  # add a row
 6  X1 = pd.concat([X1, r])  # add a row
 7  print(X1.index)
 8  X2 = X1.reset_index()
 9  print(X2.index)
10
11  X3 = X2.drop([398, 399, 400])  # drop rows
12  print(X3.tail())
```

In the above code, we create a row by selecting randomly values for the columns. Three different methods are used to add a row to the data frame. Once rows are added to a data frame, we usually need to reset the index so that the indices will be unique. Executing the above block of code gives the following output:

```
 1  Index([  0,    1,    2,    3,    4,    5,    6,    7,    8,    9,
 2          ...
 3          391, 392, 393, 394, 395, 396, 397,   0, 399,    0],
 4          dtype='int64', length=401)
 5  RangeIndex(start=0, stop=401, step=1)
 6        index  displacement  cylinders  ...  acceleration
              model_year  origin
 7  393   393          140.0         4.0  ...          15.6
              82.0      1.0
 8  394   394           97.0         4.0  ...          24.6
              82.0      2.0
 9  395   395          135.0         4.0  ...          11.6
              82.0      1.0
10  396   396          120.0         4.0  ...          18.6
              82.0      1.0
11  397   397          119.0         4.0  ...          19.4
              82.0      1.0
12
13  [5 rows x 8 columns]
```

The reset_index function does not modify the original data. Instead, it returns a copy of the data with the old index as a column.

Real data usually contain missing values. The following block of code shows how to get the rows with missing values from a data frame:

```
 1  X4 = X[X.isnull().any(axis=1)]
 2  print(X4)
```

The output of executing the above code is

```
     displacement   cylinders   horsepower   ...   acceleration
                 model_year   origin
32               98.0          4        NaN   ...           19.0
                 71        1
126             200.0          6        NaN   ...           17.0
                 74        1
330              85.0          4        NaN   ...           17.3
                 80        2
336             140.0          4        NaN   ...           14.3
                 80        1
354             100.0          4        NaN   ...           15.8
                 81        2
374             151.0          4        NaN   ...           20.5
                 82        1

[6 rows x 7 columns]
```

To fill missing values, we can use the `fillna` function. For example, we can use the following code to fill zeros for the missing values:

```
X5 = X.fillna(0)
print(X5.isna().sum())
```

Executing the above code gives the following output:

```
displacement    0
cylinders       0
horsepower      0
weight          0
acceleration    0
model_year      0
origin          0
dtype: int64
```

3.5 File IO

Pandas provides many functions for writing data to and reading data from various sources such as text files, binary files, and databases. In this section, we just illustrate how to write data frames to text files and read data from text files.

Let X and y be the features and the target of the auto MPG dataset frame from the previous section. To write the features and the target together to a CSV file, we can proceed as follows:

```
1  X["mpg"] = y
2  X.to_csv("autompg.csv")
```

Executing the above two lines of code will create a CSV file named autompg.csv in the working directory.

Reading a data frame from a CSV file can be done by using the read_csv function. The following code reads the data from the CSV file created before:

```
1  dat = pd.read_csv("autompg.csv", index_col=0)
2  print(dat.head())
3  print(dat.tail())
```

The first column is set to be the row index. The output of executing the above code is

```
1        displacement  cylinders  horsepower   ...   model_year
            origin    mpg
2  0           307.0          8       130.0   ...           70
            1    18.0
3  1           350.0          8       165.0   ...           70
            1    15.0
4  2           318.0          8       150.0   ...           70
            1    18.0
5  3           304.0          8       150.0   ...           70
            1    16.0
6  4           302.0          8       140.0   ...           70
            1    17.0
7
8  [5 rows x 8 columns]
9        displacement  cylinders  horsepower   ...   model_year
            origin    mpg
10 393         140.0          4        86.0   ...           82
            1    27.0
11 394          97.0          4        52.0   ...           82
            2    44.0
12 395         135.0          4        84.0   ...           82
            1    32.0
13 396         120.0          4        79.0   ...           82
            1    28.0
14 397         119.0          4        82.0   ...           82
            1    31.0
15
16 [5 rows x 8 columns]
```

3.6 Summary

Pandas is an important Python packages for working with tabular data. In this chapter, we gave a brief introduction to the Pandas package. In particular, we introduced Pandas data structures and how to manipulate these data structures. For a complete reference of functions provided by Pandas, readers are referred to the website of the package at `https://pandas.pydata.org`.

4

The Matplotlib Library

The Matplotlib library is a comprehensive Python library that is used to create visualizations. It is built on NumPy and can be used to create static, animated, and interactive visualizations. In this chapter, we shall give a brief introduction to the Matplotlib library.

4.1 Overview

In Matplotlib, creating a figure involves manipulating the following objects: a `Figure` object, `Axes` objects, `Axis` objects, and `Artist` objects. Figure 4.1 shows a hierarchical diagram of a few selected classes of Matplotlib. The `Figure` object is the top container of other objects. However, the `Figure` object is also an `Artist` object. In fact, almost all objects in Matplotlib are `Artist` objects.

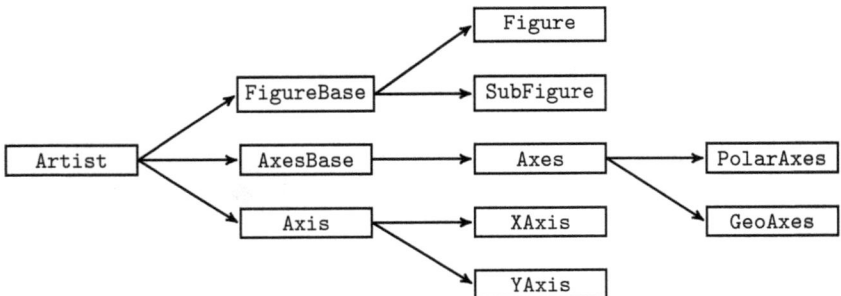

FIGURE 4.1: A hierarchical diagram of selected classes of Matplotlib.

An `Axes` object is used to configure a plotting area and includes two or three `Axis` objects. An `Axis` object is used to specify ticks, tick labels, and scales for the data in the `Axes` object. Most parts of a figure are configured by using methods from the `Axes` class.

The inputs to Matplotlib plotting functions are NumPy arrays or objects that can be converted to NumPy arrays by the NumPy function `asarray`.

DOI: 10.1201/9781003592648-4   66

Pandas data objects and NumPy matrices may not work for Matplotlib. These data objects need to be converted to NumPy arrays before being plotted.

Matplotlib supports the following two major coding styles:

Object-oriented style This coding style explicitly creates `Figure` objects and `Axes` objects, and controls them by using methods defined on them.

Matlab style This coding style resembles plotting figures in Matlab, which is a proprietary software for scientific computing. This style relies on functions in the `pyplot` module to automatically create and manage figures.

The following block of code illustrates the two coding style:

```python
import numpy as np
import matplotlib.pyplot as plt

t = np.arange(1, 101)
y = np.cumsum(np.random.standard_normal(100))

# OO style
fig, ax = plt.subplots(figsize=(5, 3))
ax.plot(t, y, color="black")
ax.set_xlabel('t')
ax.set_ylabel('y')
ax.set_title("Random walk")

# Matlab style
plt.figure(figsize=(5, 3))
plt.plot(t, y, color="black")
plt.xlabel('t')
plt.ylabel('y')
plt.title("Random walk")
```

In the above code, we first import the NumPy package and the Matplotlib package. Then we use NumPy functions to generate a sequence of time steps and a random walk. In the object-oriented style, the `subplots` function is used to create a `Figure` object and an `Axes` object. Then the `Axes` object is manipulated to display the data. In the Matlab style, functions from the `pyplot` module are used to create the plot. The two styles will create exactly the same figure, which is shown in Figure 4.2.

The differences of the two coding styles have been documented in Matplotlib's documentation, which is available at https://matplotlib.org/. For creating simple plots, both coding styles can be used. For creating complex plots that involves many objects, the object-oriented style is better than the Matlab style as the former allows easy references to the plotting objects. In this book, we will use the object-oriented style to plot figures.

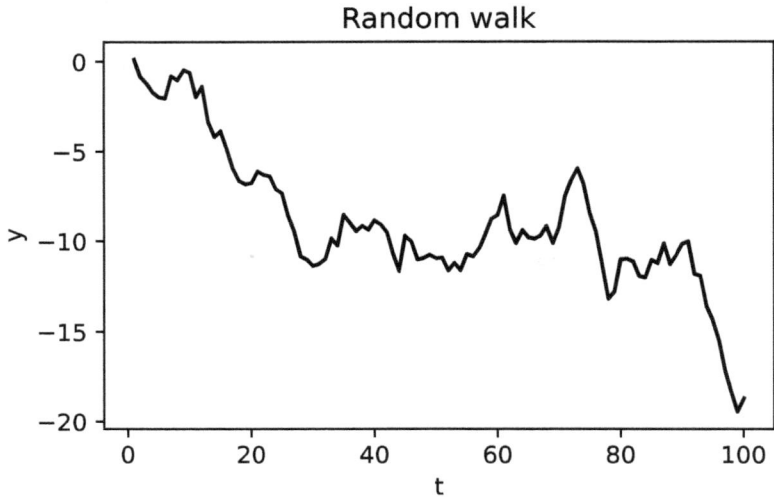

FIGURE 4.2: A simple random walk created by Matplotlib.

4.2   Basic plotting

In this section, we illustrate how to use Matplotlib to create basic plots such
as scatter plots, box plots, and histograms.

A scatter plot is used to plot a pair of two variables. It is used to examine
the relationship between two variables. In Matplotlib, we can use the **scatter**
function to create scatter plots. The following block of code illustrates how to
create scatter plots in Matplotlib:

```
from ucimlrepo import fetch_ucirepo

auto_mpg = fetch_ucirepo(id=9)

X = auto_mpg.data.features
y = auto_mpg.data.targets

fig1, ax1 = plt.subplots(figsize=(4, 3))
ax1.scatter(X["cylinders"], y, color="black", s=8)
ax1.set_xlabel('Cylinders')
ax1.set_ylabel('mpg')

fig2, ax2 = plt.subplots(figsize=(4, 3))
ax2.scatter(X["horsepower"], y, color="black", s=8)
ax2.set_xlabel('Horse power')
ax2.set_ylabel('mpg')
```

In the above code, we first get the auto MPG dataset from the UCI machine learning repository. Then we create a scatter plot between the variables `cylinders` and `mpg`. Afterwards, we create a scatter plot between the variables `horsepower` and `mpg`. The argument `s=8` used in the `scatter` function specifies the size of the points. The results scatter plots are shown in Figure 4.3.

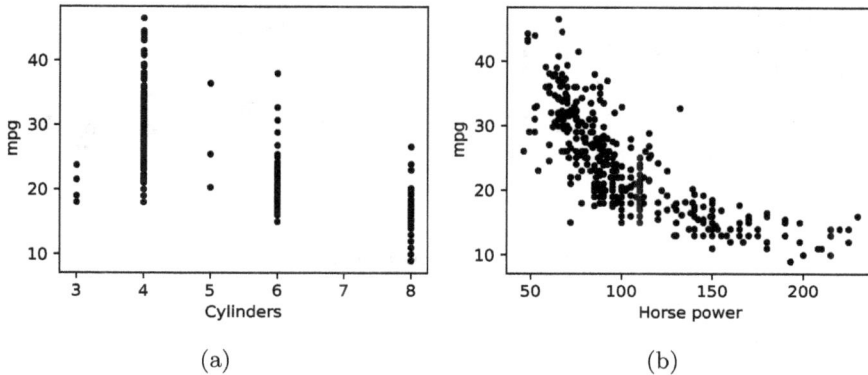

(a)   (b)

FIGURE 4.3: Two scatter plots.

A histogram is used to display the distribution of a numerical variable. It is created by dividing the data into a specified number of bins and displaying the frequencies (in count or in percentage) of the data falling in the bins. The following block of code illustrates how to create histograms in Matplotlib:

```
from matplotlib.ticker import PercentFormatter

fig3, ax3 = plt.subplots(figsize=(4, 3))
ax3.hist(X["displacement"], bins=50, color="black")

fig4, ax4 = plt.subplots(figsize=(4, 3))
ax4.hist(y, bins=50, color="black", density=True)
ax4.yaxis.set_major_formatter(PercentFormatter(xmax=1))
```

The above code creates two histograms. In the first histogram, the counts are displayed in the vertical axis. In the second histogram, the percentages are displayed in the vertical axis. The tick labels of the vertical axis of the second histogram are formatted by the function `PercentFormatter` from the `ticker` module. The resulting histograms are shown in Figure 4.4.

Like a histogram, a box plot is also used to display the distribution of a quantitative variable. A box plot displays the interquantile range of the data and shows the least and greatest values. In Matplotlib, the `boxplot` function is used to create box plots. The following block of code creates two box plots:

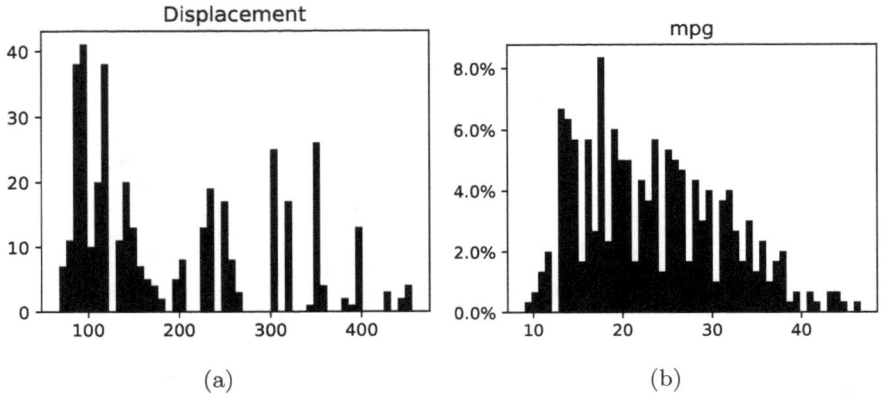

FIGURE 4.4: Two histograms.

```
fig5, ax5 = plt.subplots(figsize=(4, 3))
ax5.boxplot(X["displacement"], patch_artist=True,
            boxprops=dict(color='black', facecolor='
                white'),
            whiskerprops=dict(color='black'),
            capprops=dict(color='black'),
            medianprops=dict(color='black'),
            flierprops=dict(marker='o', color='black',
                markersize=5))
ax5.set_title("Displacement")

fig6, ax6 = plt.subplots(figsize=(4, 3))
ax6.boxplot(y, patch_artist=True,
            boxprops=dict(color='black', facecolor='
                white'),
            whiskerprops=dict(color='black'),
            capprops=dict(color='black'),
            medianprops=dict(color='black'),
            flierprops=dict(marker='o', color='black',
                markersize=5))
ax6.set_title("mpg")
```

In the above code, we create box plots for two variables of the auto MPG dataset. All components of a box plot can be customized. The resulting box plots are shown in Figure 4.5.

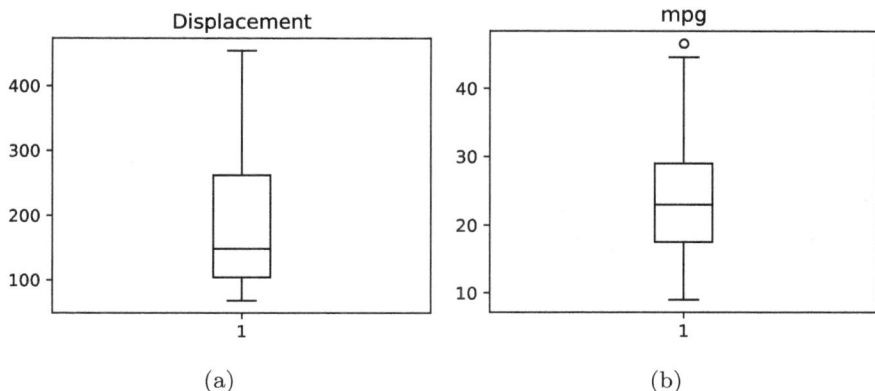

FIGURE 4.5: Two histograms.

4.3 Subplots

Subplots are plots that appear in a figure together. Producing subplots in Matplotlib is straightforward. The following block of code illustrates how to create subplots in Matplotlib:

```
iris = fetch_ucirepo(id=53)

X = iris.data.features
y = iris.data.targets

varnames = list(X.columns)
fig7, ax7 = plt.subplots(3,2, figsize=(6,9))
fig7.tight_layout()
nCount = 0;
for i in range(3):
    for j in range(i+1, 4):
        rowInd = nCount // 2
        colInd = nCount - 2*rowInd
        nCount += 1
        ax7[rowInd][colInd].scatter(X[varnames[i]], X[
            varnames[j]], color='black', s=5)
        ax7[rowInd][colInd].set_xlabel(varnames[i])
        ax7[rowInd][colInd].set_ylabel(varnames[j])
```

In the above code, we first load the Iris dataset, which has an id of 53 in the UCI machine learning repository. Then we create a figure with 3 subplots. Then we use nested `for` loops to add scatter plots to these subplots. The `tight_layout` function is used to automatically adjust the spacing between subplots to prevent overlap. The resulting figure is shown in Figure 4.6.

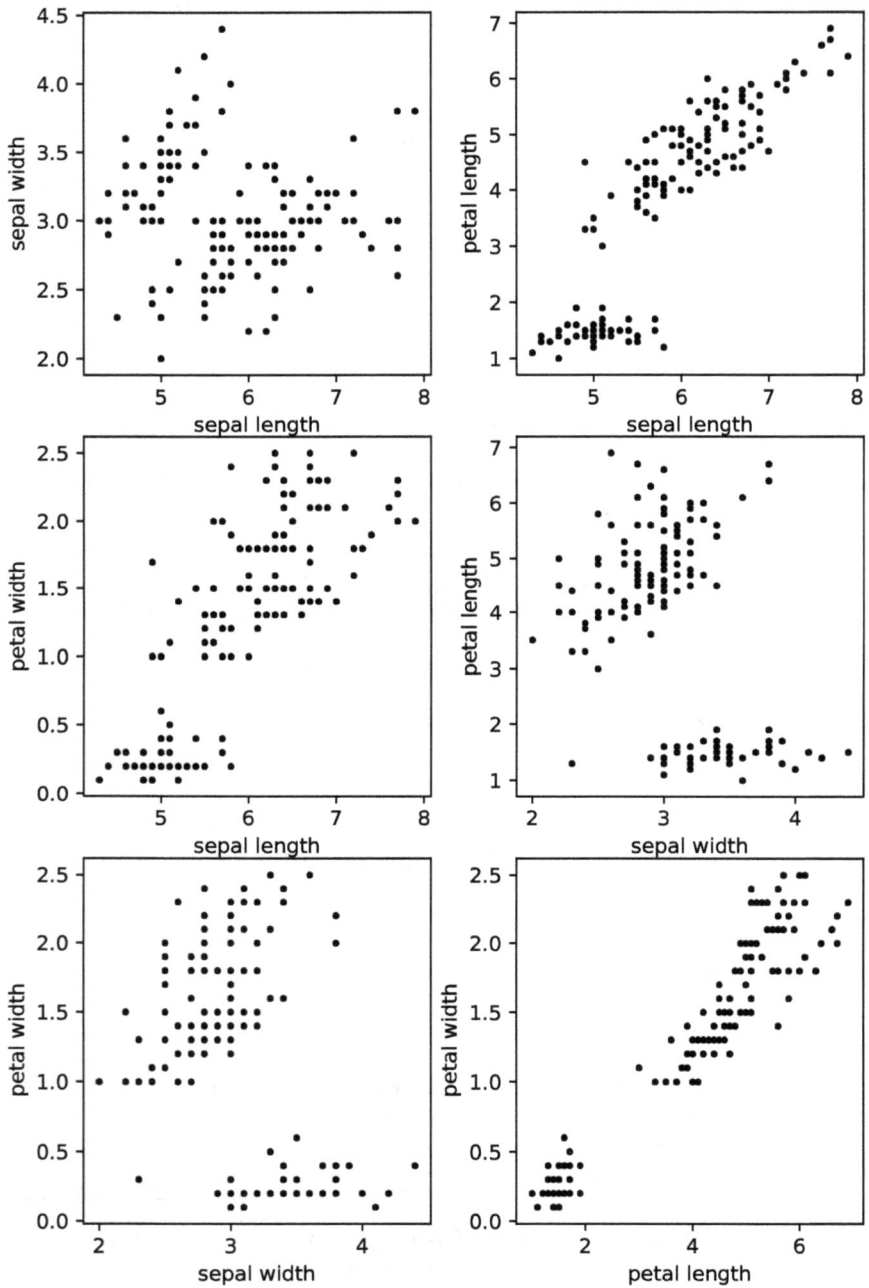

FIGURE 4.6: Scatter plots of the Iris dataset.

4.4 File IO

Matplotlib provides functions to save plots to files. This feature is extremely useful when executing Python code in servers. The following code illustrates how to save a figure to a PDF file:

```
fig7.savefig("iris.pdf", bbox_inches='tight')
```

In the above code, the object `fig7` is the figure object created in the previous section. Executing the above line of code will produce a PDF file in the working directory. In addition to PDF files, Matplotlib can also save figures to other file formats.

4.5 Summary

In this chapter, we gave a brief introduction to the Matplotlib package, which is an important Python package for visualizing data. We introduced some functions for creating basic plots and subplots. Matplotlib supports many other types of plots. For a complete guide, readers are referred to the package's documentation at `https://matplotlib.org/`.

Part II

Data Clustering in Python

5

Introduction to Data Clustering

Data clustering is a fundamental data mining technique that groups similar data points together, enabling the discovery of patterns, identification of trends, and facilitation of data-driven decision-making. In this chapter, we provide a concise introduction to the key concepts and methods of data clustering.

5.1 History of Clustering

Data clustering, or cluster analysis, refers to a process of dividing a set of objects into clusters, which are homogeneous groups that are externally isolated and internally cohesive. Data clustering emerged as a major research topic in the 1960s and the sum-of-squares criterion was used to develop clustering algorithms. The k-means algorithm is a clustering algorithm based on the sum-of-squares criterion and has been proposed by several researchers in different forms and under different assumptions.

The sum-of-squares criterion comes in two versions [28, 29]: the discrete version and the continuous version. Let $X = \{\mathbf{x}_1, \mathbf{x}_2, \ldots, \mathbf{x}_n\}$ denote a set of n data points in a d-dimensional space. Let k denote the desired number of clusters. The discrete version of the sum-of-squares criterion is formulated as follows:

$$L(Z) = \sum_{l=1}^{k} \sum_{\mathbf{x} \in C_l} \|\mathbf{x} - \mathbf{z}_l\|^2, \tag{5.1}$$

where $\mathcal{C} = \{C_1, C_2, \ldots, C_k\}$ is the set of clusters, $Z = \{\mathbf{z}_1, \mathbf{z}_2, \ldots, \mathbf{z}_k\}$ is the set of cluster centers, and $\mathbf{z}_l$ is the center (also called the centroid or the prototype) of the lth cluster. Here $\| \cdot \|$ denotes the L^2 norm or Euclidean distance, which is defined as:

$$\|\mathbf{x} - \mathbf{y}\| = \sqrt{\sum_{j=1}^{d} (x_j - y_j)^2}, \tag{5.2}$$

DOI: 10.1201/9781003592648-5

where $\mathbf{x}$ and $\mathbf{y}$ represent two data points.

The continuous version of the sum-of-squares criterion is formulated as follows [29]:

$$L(\mathcal{C}) = \sum_{l=1}^{k} \int_{C_l} \|\mathbf{x} - E[\mathbf{X}|\mathbf{X} \in C_l]\|^2 \, d P(\mathbf{x}), \qquad (5.3)$$

where $\mathbf{X}$ a random vector with a distribution P in $\mathbb{R}^d$ and $\mathcal{C} = \{C_1, C_2, \ldots, C_k\}$ is a partition of $\mathbb{R}^d$. However, it might be computationally challenging to calculate the cluster centers under the continuous sum-of-squares criterion.

Many methods have been proposed to minimize the discrete criterion. These methods can be classified into two types: exact minimization and approximate minimization. Methods for exact minimization include integer programming [212, 245], dynamic programming [19, 150], branch-and-bound methods [188], and cutting plane methods [114]. Exact minimization methods are impractical for large clustering problems. Approximate minimization methods have been developed to handle large clustering problems. Approximate methods include traditional heuristics (e.g., k-means) and metaheuristics (e.g., simulated annealing, Tabu search, genetic algorithm).

In the 1950s, several researchers studied the use of the sum-of-squares criteria for clustering [47, 50, 51, 80, 236]. Dalenius and Gurney [51, 50] were among the first to formulate the clustering problem under the continuous sum-of-squares criterion, although they did not use a k-means algorithm to do the minimization. Cox [47] defined the average loss from grouping as follows:

$$L = \frac{E\left[(Y - E[Y])^2\right]}{\sigma^2} = \frac{1}{\sigma^2} \sum_{l=1}^{k} p_l E\left[(Y - \xi_l)^2 | Y \in C_l\right]$$

$$= 1 - \frac{1}{\sigma^2} \sum_{l=1}^{k} p_l \left(\xi_l - E[Y]\right)^2, \qquad (5.4)$$

where Y is a random variable with standard deviation σ, k denotes the number of groups, C_l is the lth group, ξ_l is the mean of all observations in C_l, and p_l is the probability of an observation falling in C_l. Cox [47] also considered a special case when X follows a normal distribution. For $k = 3$, the three groups are $(-\infty, -y)$, $(-y, y)$, and (y, ∞), where y is chosen such that the average loss is minimized. In this case, the average loss is

$$L = 1 - \frac{2\varphi(y)^2}{\Phi(-y)},$$

where $\varphi(\cdot)$ and $\Phi(\cdot)$ are the probability density function and the cumulative distribution function of the standard normal distribution, respectively. The average loss reaches the minimum when $y = 0.612$.

Fisher [80] investigated the clustering problem under the discrete sum-of-squares criterion. In particular, Fisher [80] proposed a practical procedure to group a set of arbitrary numbers so that the variance within groups is minimized. Given n numbers $x_1, x_2, \ldots, x_n$. For $i = 1, 2, \ldots, n$, the ith number is associated with a weight w_i. The variance within groups of dividing these n numbers into k groups is defined as

$$D = \sum_{i=1}^{n} w_i(x_i - z_{\gamma(i)})^2, \tag{5.5}$$

where z_l denotes the weighted arithmetic mean of numbers in the lth group and $\gamma(i)$ denotes the index of the group to which x_i is assigned. Fisher [80] showed that among all possible partitions, only contiguous partitions need to be considered. A contiguous partition is one in which, for any numbers x_i, x_j, and x_s satisfying $x_i < x_j < x_s$, if x_i and x_s belong to the same group, then x_j must also belong to that group.

Regarding the origins of the k-means algorithm, Steinley [237] noted that Sebestyen [228] and MacQueen [174] independently developed it as a strategy for minimizing the sum-of-squares criterion. Bock [29] pointed out that Steinhaus [236] first proposed explicitly the continuous k-means algorithm in the multidimensional case, Forgy [83] first proposed the discrete k-means algorithm, and MacQueen [174] first used the name "k-means algorithm." In addition to propose the continuous version of the k-means algorithm, Steinhaus [236] discussed the existence and uniqueness of a solution. Forgy [83] published an abstract of his presentation at the Spring meeting of ENAR (Eastern North American Region) held at the Florida State University at Tallahassee on April 29 to May 1, 1965. The k-means algorithm was not included in the abstract. However, Anderberg [10, p161] and MacQueen [174, p294] described the details of Forgy's presentation. In addition, Lloyd studied the continuous sum-of-squares criterion in $\mathbb{R}^1$ and proposed an one-dimensional version of the k-means algorithm in 1957. The algorithm was not published in a journal until 1982 (see [170]).

Since the k-means algorithm was proposed in the 1960s, data clustering has become an increasingly popular research topic. Figure 5.1 shows the number of publications related to data clustering from 1976 to 2024. The titles or the abstracts of these publications contain the keyword "data clustering." In reality, the number of publications should be much more than shown in the figure because only digitized publications are counted. From the figure, we see that publications related to data clustering show an exponentially increase trend over the past 50 years.

5.2   Data Clustering Process

A typical clustering process involves the following five steps [144]:

(a) pattern representation;

(b) dissimilarity measure definition;

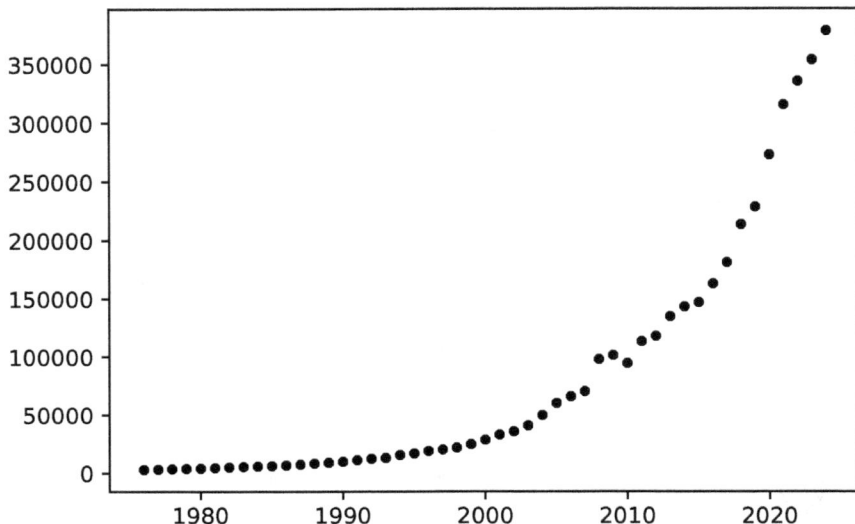

FIGURE 5.1: Publications related to data clustering from 1976 to 2024. Data source: app.dimensions.ai.

(c) clustering;

(d) data abstraction;

(e) assessment of output.

In the pattern representation step, the number and type of the attributes are determined. Feature selection, the process of identifying the most effective subset of the original attributes to use in clustering, and feature extraction, the process of transforming the original attributes to new attributes, are also done in this step if needed.

In the dissimilarity measure definition step, a distance measure appropriate to the data domain is defined. Various distance measures have been developed and used in data clustering [90]. The most common one among them, for example, is the Euclidean distance.

In the clustering step, a clustering algorithm is used to group a set of records into a number of meaningful clusters. The clustering can be hard clustering, where each record belongs to one and only one cluster, or fuzzy clustering, where a record can belong to two or more clusters with probabilities. The clustering algorithm can be hierarchical, where a nested series of partitions is produced, or partitional, where a single partition is identified.

In the data abstraction step, one or more prototypes (i.e., representative records) of a cluster is extracted so that the clustering results are easy to comprehend. For example, a cluster can be represented by a centroid.

In the final step, the output of a clustering algorithm is assessed. There are three types of assessments: external, internal, and relative [146]. In an external assessment, the recovered structure of the data is compared to the a priori structure. In an internal assessment, one tries to determine whether the structure is intrinsically appropriate to the data. In a relative assessment, a test is performed to compare two structures and measure their relative merits.

5.3   Clusters

Over the last 50 years, thousands of clustering algorithms have been developed [145]. However, there is still no formal uniform definition of the term cluster. In fact, formally defining cluster is difficult and may be misplaced [72].

Although no formal definition of cluster exists, there are several operational definitions of cluster. For example, Bock [26] suggested that a cluster is a group of data points satisfying various plausible criteria such as:

(a) Share the same or closely related properties;

(b) Show small mutual distances;

(c) Have "contacts" or "relations" with at least one other data point in the group;

(d) Can be clearly distinguishable from the rest of the data points in the dataset.

Carmichael [37] suggested that a set of data points forms a cluster if the distribution of the set of data points satisfies the following conditions:

(a) Continuous and relatively dense regions exist in the data space; and

(b) Continuous and relatively empty regions exist in the data space.

Lorr [171] suggested that there are two kinds of clusters for numerical data: compact clusters and chained clusters. A compact cluster is formed by a group of data points that have high mutual similarity. For example, Figure 5.2 shows a two-dimensional dataset with three compact clusters. Usually, such a compact cluster has a center [181].

A chained cluster is formed by a group of data points in which any two data points in the cluster are reachable through a path. For example, Figures 5.3 shows a dataset with three chained clusters. Unlike a compact cluster, which

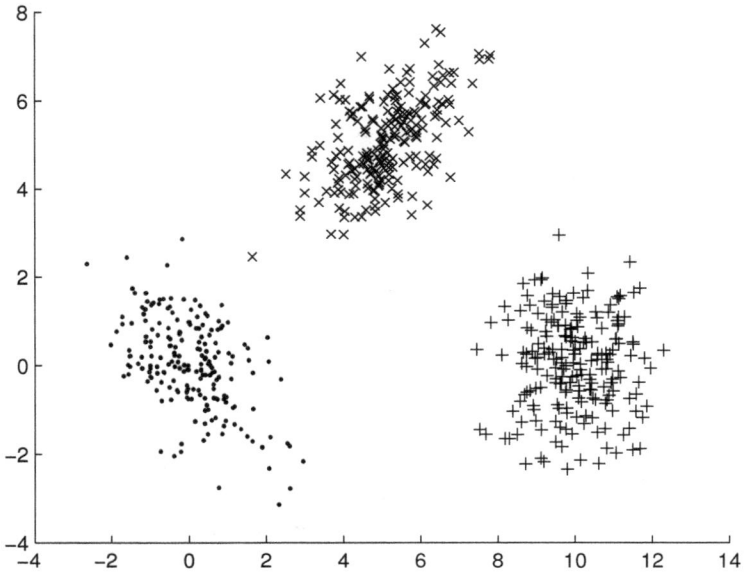

FIGURE 5.2: A dataset with three compact clusters.

can be represented by a single center, a chained cluster is usually represented by multiple centers.

Everitt [73] also summarized several operational definitions of cluster. For example, one definition of cluster is that a cluster is a set of data points which are alike and data points from different clusters are not alike. Another definition of cluster is that a cluster is a set of data points such that the distance between any two points in the cluster is less than the distance between any point in the cluster and any point not in it.

5.4 Data Types

Most clustering algorithms are associated with data types. It is important to understand different types of data in order to perform cluster analysis. By data type we mean hereby the type of a single attribute.

In terms of how the values are obtained, an attribute can be typed as discrete and continuous. The values of a discrete attribute are usually obtained by some sort of counting; while the values of a continuous attribute are

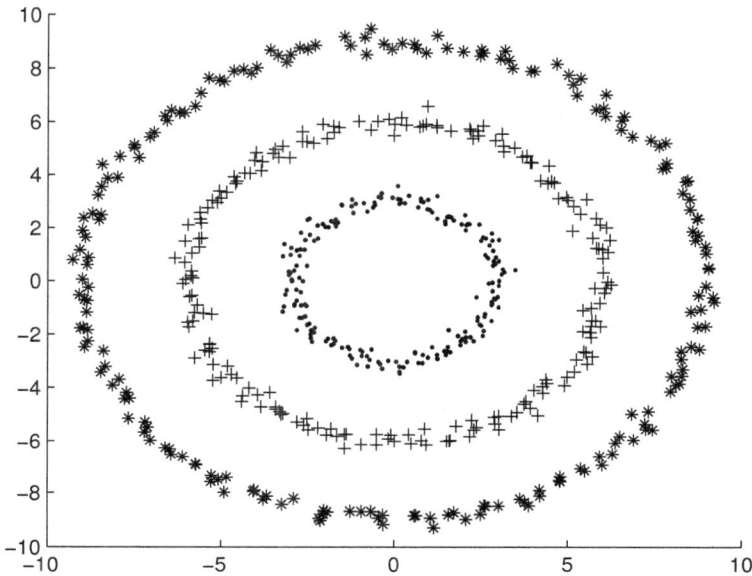

FIGURE 5.3: A dataset with three chained clusters.

obtained by some sort of measuring. For example, the number of cars is discrete and the weight of a person is continuous. There is a gap between two different discrete values and there is always a value between two different continuous values.

In terms of measurement scales, an attribute can be typed as ratio, interval, ordinal, or nominal. Nominal data is discrete data without a natural ordering. For example, name of a person is nominal. Ordinal data is discrete data that have a natural ordering. For example, the order of persons in a line is ordinal. Interval data is continuous data that have a specific order and equal intervals. For example, temperature is interval data. Ratio data is continuous data that is interval data and has a natural zero. For example, the annual salary of a person is ratio data. The ratio and interval types are continuous types; while the ordinal and nominal types are discrete types (see Table 5.1).

TABLE 5.1: Attribute types.

Continuous	Discrete
Ratio	Ordinal
Interval	Nominal

5.5   Dissimilarity and Similarity Measures

Dissimilarity or distance is an important part of clustering as almost all clustering algorithms rely on some distance measure to define the clustering criteria. Since records might have different types of attributes, the appropriate distance measures are also different. For example, the most popular Euclidean distance is used to measure dissimilarities between continuous records, i.e., records consist of continuous attributes.

A distance function D on a dataset X is a binary function that satisfies the following conditions [10, 260]:

(a) $D(\mathbf{x}, \mathbf{y}) \geq 0$ (Nonnegativity);

(b) $D(\mathbf{x}, \mathbf{y}) = D(\mathbf{y}, \mathbf{x})$ (Symmetry or Commutativity);

(c) $D(\mathbf{x}, \mathbf{y}) = 0$ if and only if $\mathbf{x} = \mathbf{y}$ (Reflexivity);

(d) $D(\mathbf{x}, \mathbf{y}) \leq D(\mathbf{x}, \mathbf{z}) + D(\mathbf{z} + \mathbf{y})$ (Triangle inequality),

where $\mathbf{x}$, $\mathbf{y}$, and $\mathbf{z}$ are arbitrary data points in X. A distance function is also called a metric, which satisfies the above four conditions.

If a function satisfies the first three conditions and does not satisfy the triangle inequality, then the function is called a semimetric. In addition, if a metric D satisfies the following condition

$$D(\mathbf{x}, \mathbf{y}) \leq \max\{D(\mathbf{x}, \mathbf{z}), D(\mathbf{z} + \mathbf{y})\},$$

then the metric is called an ultrametric [153].

Unlike distance measures, similarity measures are defined in the opposite way. The more the two data points similar to each other, the larger the similarity is and the smaller the distance is.

5.5.1   Measures for Continuous Data

The most common distance measure for continuous data is the Euclidean distance. Given two data points $\mathbf{x}$ and $\mathbf{y}$ in a d-dimensional space, the Euclidean distance between the two data points is defined as

$$D_{euc}(\mathbf{x}, \mathbf{y}) = \sqrt{\sum_{j=1}^{d}(x_j - y_j)^2}, \tag{5.6}$$

where x_j and y_j are the jth components of $\mathbf{x}$ and $\mathbf{y}$, respectively.

The Euclidean distance measure is a metric [260]. Clustering algorithms that use the Euclidean distance tend to produce hyperspherical clusters. Clusters produced by clustering algorithms that use the Euclidean distance are invariant to translations and rotations in the data space [65]. One disadvantage of the Euclidean distance is that attributes with large values and variances dominate other attributes with small values and variances. However, this problem can be alleviated by normalizing the data so that each attribute contributes equally to the distance.

The squared Euclidean distance between two data points is defined as

$$D_{squ}(\mathbf{x}, \mathbf{y}) = \sum_{j=1}^{d}(x_j - y_j)^2. \tag{5.7}$$

The Manhattan distance or city block distance between two data points is defined as

$$D_{man}(\mathbf{x}, \mathbf{y}) = \sum_{j=1}^{d}|x_j - y_j|. \tag{5.8}$$

The maximum distance between two data points is defined as

$$D_{max}(\mathbf{x}, \mathbf{y}) = \max_{1 \le j \le d}|x_j - y_j|. \tag{5.9}$$

The Euclidean distance and the Manhattan distance are special cases of the Minkowski distance, which is defined as

$$D_{min}(\mathbf{x}, \mathbf{y}) = \left(\sum_{j=1}^{d}|x_j - y_j|^p\right)^{\frac{1}{p}}, \tag{5.10}$$

where $p \ge 1$. In fact, the maximum distance is also a special case of the Minkowski distance when we let $p \to \infty$.

The Mahalanobis distance is defined as

$$D_{mah}(\mathbf{x}, \mathbf{y}) = \sqrt{(\mathbf{x} - \mathbf{y})^T \Sigma^{-1} (\mathbf{x} - \mathbf{y})}, \tag{5.11}$$

where Σ^{-1} is the inverse of a covariance matrix Σ, $\mathbf{x}$ and $\mathbf{y}$ are column vectors, $(\mathbf{x} - \mathbf{y})^T$ denotes the transpose of $(\mathbf{x} - \mathbf{y})$. The Mahalanobis distance can be used to alleviate the distance distortion caused by linear combinations of attributes [146, 176].

Some other distance measures for continuous data have also been proposed. For example, the average distance [163], the generalized Mahalanobis distance [186], the weighted Manhattan distance [255], the chord distance [200], and the Pearson correlation [69], to name just a few. Many other distance measures for numeric data can be found in [90].

5.5.2 Measures for Discrete Data

The most common distance measure for discrete data is the simple matching distance. The simple matching distance between two categorical data points

x and **y** is defined as [134, 135, 136, 156]:

$$D_{sim}(\mathbf{x}, \mathbf{y}) = \sum_{j=1}^{d} \delta(x_j, y_j), \qquad (5.12)$$

where d is the dimension of the data points and $\delta(\cdot, \cdot)$ is defined as

$$\delta(x_j, y_j) = \begin{cases} 0 & \text{if } x_j = y_j, \\ 1 & \text{if } x_j \neq y_j. \end{cases}$$

Some other matching coefficients for categorical data have also been proposed. For a comprehensive list of matching coefficients, readers are referred to [90, Chapter 6]. A comprehensive list of similarity measures for binary data, which is a special case of categorical data, can also be found in [90].

5.5.3 Measures for Mixed-type Data

A dataset might contain both continuous and discrete data. In this case, we need to use measure for mixed-type data. Gower [109] proposed a general similarity coefficient for mixed-type data, which is defined as

$$S_{gow}(\mathbf{x}, \mathbf{y}) = \frac{1}{\sum\limits_{j=1}^{d} w(x_j, y_j)} \sum_{j=1}^{d} w(x_j, y_j) s(x_j, y_j), \qquad (5.13)$$

where $s(x_j, y_j)$ is a similarity component for the jth components of **x** and **y**, and $w(x_j, y_j)$ is either one or zero depending on whether a comparison for the jth component of the two data points is valid or not.

For different types of attributes, $s(x_j, y_j)$ and $w(x_j, y_j)$ are defined differently. If the jth attribute is continuous, then

$$s(x_j, y_j) = 1 - \frac{|x_j - y_j|}{R_j},$$

$$w(x_j, y_j) = \begin{cases} 0 & \text{if } x_j \text{ or } y_j \text{ is missing}, \\ 1 & \text{otherwise}, \end{cases}$$

where R_j is the range of the jth attribute.

If the jth attribute is binary, then

$$s(x_j, y_j) = \begin{cases} 1 & \text{if both } x_j \text{ and } y_j \text{ are "present"}, \\ 0 & \text{otherwise}, \end{cases}$$

$$w(x_j, y_j) = \begin{cases} 0 & \text{if both } x_j \text{ and } y_j \text{ are "absent"}, \\ 1 & \text{otherwise}. \end{cases}$$

If the jth attribute is nominal or categorical, then

$$s(x_j, y_j) = \left\{ \begin{array}{ll} 1 & \text{if } x_j = y_j, \\ 0 & \text{otherwise}, \end{array} \right.$$

$$w(x_j, y_j) = \left\{ \begin{array}{ll} 0 & \text{if } x_j \text{ or } y_j \text{ is missing}, \\ 1 & \text{otherwise}. \end{array} \right.$$

A general distance measure was defined similarly in [109]. Ichino and Yaguchi [138, 139] proposed a generalized Minkowski distance, which was also presented in [90].

5.6 Hierarchical Clustering Algorithms

A hierarchical clustering algorithm is a clustering algorithm that divides a dataset into a sequence of nested partitions. Hierarchical clustering algorithms can be further classified into two categories: agglomerative hierarchical clustering algorithms and divisive hierarchical clustering algorithms.

An agglomerative hierarchical algorithm starts with every single record as a cluster and then repeats merging the closest pair of clusters according to some similarity criteria until only one cluster is left. For example, Figure 5.4 shows an agglomerative clustering of a dataset with 5 records.

In contrast to agglomerative clustering algorithms, a divisive clustering algorithm starts with all records in a single cluster and then repeats splitting large clusters into smaller ones until every cluster contains only a single record. Figure 5.5 shows an example of divisive clustering of a dataset with 5 records.

5.6.1 Agglomerative Hierarchical Algorithms

Based on different ways to calculate the distance between two clusters, agglomerative hierarchical clustering algorithms can be classified into the following several categories [189]:

(a) Single linkage algorithms;

(b) Complete linkage algorithms;

(c) Group average algorithms;

(d) Weighted group average algorithms;

(e) Ward's algorithms;

(f) Centroid algorithms;

(g) Median algorithms;

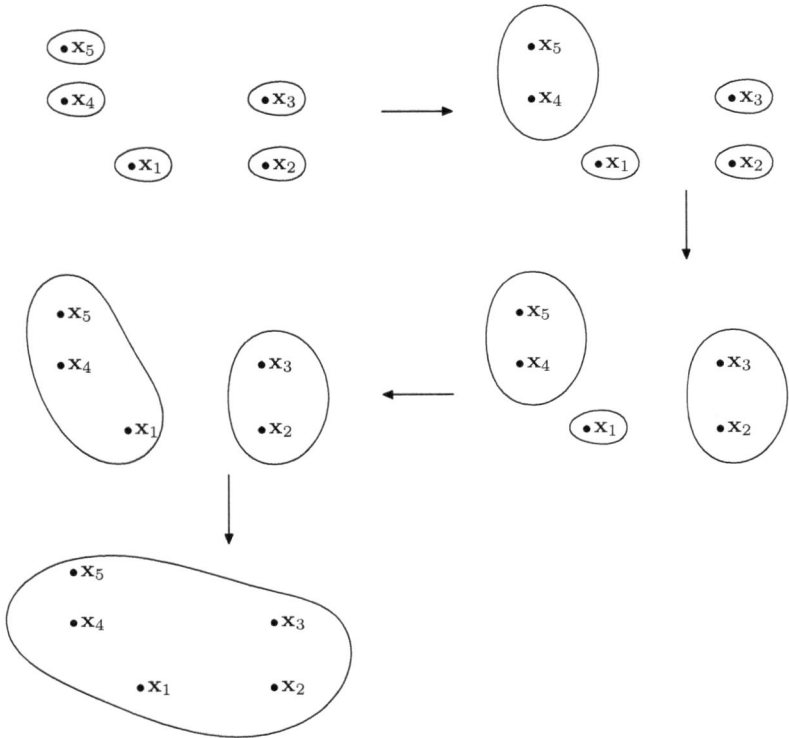

FIGURE 5.4: Agglomerative clustering.

(h) Other agglomerative algorithms that do not fit into the above categories.

For algorithms in the first seven categories, we can use the Lance-Williams recurrence formula [161, 162] to calculate the distance between an existing cluster and a cluster formed by merging two existing clusters. The Lance-Williams formula is defined as

$$
\begin{aligned}
& D(C_k, C_i \cup C_j) \\
= \ & \alpha_i D(C_k, C_i) + \alpha_j D(C_k, C_j) \\
& + \beta D(C_i, C_j) + \gamma |D(C_k, C_i) - D(C_k, C_j)|,
\end{aligned}
$$

where C_k, C_i, and C_j are three clusters, $C_i \cup C_j$ denotes the cluster formed by merging clusters C_i and C_j, $D(\cdot, \cdot)$ is a distance between clusters, and α_i, α_j, β, and γ are adjustable parameters. Section 6.1 presents various values of these parameters.

When the Lance-Williams formula is used to calculate distances, the single linkage and the complete linkage algorithms induce a metric on the dataset

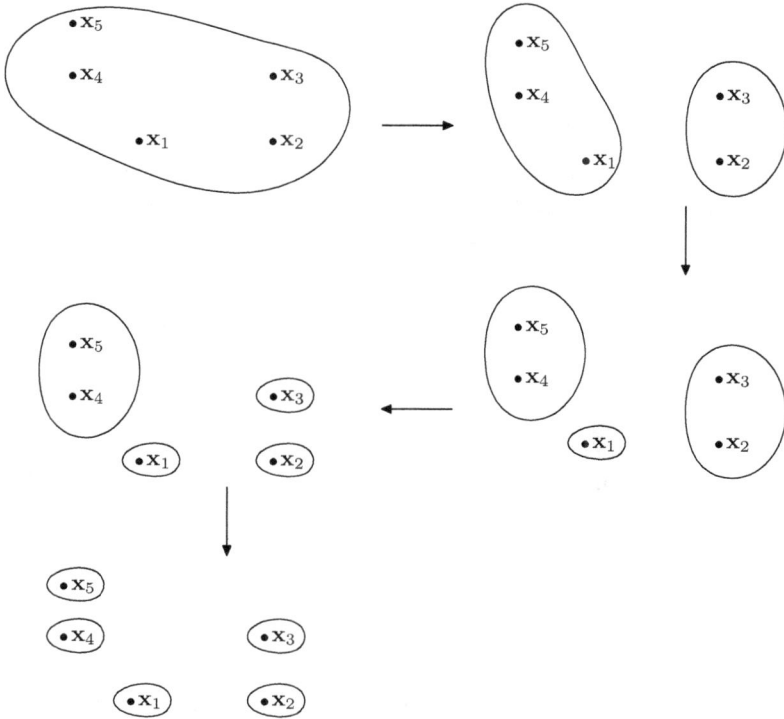

FIGURE 5.5: Divisive clustering.

known as the ultrametric [153]. However, other agglomerative algorithms that use the Lance-Williams formula might not produce a ultrametric [182].

A more general recurrence formula has been proposed in [148] and discussed in [107] and [90]. The general recurrence formula is defined as

$$
\begin{aligned}
& D(C_k, C_i \cup C_j) \\
= \ & \alpha_i D(C_k, C_i) + \alpha_j D(C_k, C_j) \\
& + \beta D(C_i, C_j) + \gamma |D(C_k, C_i) - D(C_k, C_j)| \\
& + \delta_i h(C_i) + \delta_j h(C_j) + \epsilon h(C_k),
\end{aligned}
$$

where $h(C)$ denotes the height of cluster C in the dendrogram, and δ_i, δ_j, and ϵ are adjustable parameters. Other symbols are the same as in the Lance-Williams formula. If we let the three parameters δ_i, δ_j, and ϵ be zeros, then the general formula becomes the Lance-Williams formula.

Some other agglomerative hierarchical clustering algorithms are based on the general recurrence formula. For example, the flexible algorithms [161], the sum of squares algorithms [148], and the mean dissimilarity algorithms [107, 129] are such agglomerative hierarchical clustering algorithms.

5.6.2 Divisive Hierarchical Algorithms

Divisive hierarchical algorithms can be classified into two categories: monothetic and polythetic [73, 254]. A monothetic algorithm divides a dataset based on a single specified attribute. A polythetic algorithm divides a dataset based on the values of all attributes.

Given a dataset containing n records, there are $2^n - 1$ nontrivial different ways to split the dataset into two pieces [68]. As a result, it is not feasible to enumerate all possible ways of dividing a large dataset. Another difficulty of divisive hierarchical clustering is to choose which cluster to split in order to ensure monotonicity.

Divisive hierarchical algorithms that do not consider all possible divisions and that are monotonic do exist. For example, the algorithm DIANA (DIvisive ANAlysis) is such a divisive hierarchical clustering algorithm [156].

5.6.3 Other Hierarchical Algorithms

In the previous two subsections, we presented several classic hierarchical clustering algorithms. These classic hierarchical clustering algorithms have drawbacks. One drawback of these algorithms is that they are sensitive to noise and outliers. Another drawback of these algorithms is that they can not handle large datasets since their computational complexity is at least $O(n^2)$ [260], where n is the size of the dataset. Several hierarchical clustering algorithms have been developed in an attempt to improve these drawbacks. For example, BIRCH [273], CURE [117], ROCK [115], and Chameleon [155] are such hierarchical clustering algorithms.

Other hierarchical clustering algorithms have also been developed. For example, [164] proposed an agglomerative hierarchical clustering algorithm based on the scale-space theory in human visual system research. [166] proposed a similarity-based agglomerative clustering (SBAC) to cluster mixed-type data. [17] proposed a divisive hierarchical clustering algorithm based on unsupervised decision trees.

5.6.4 Dendrograms

Results of a hierarchical clustering algorithm are usually visualized by dendrograms. A dendrogram is a tree in which each internal node is associated with a height. The heights in a dendrogram satisfy the following ultrametric conditions [153]:

$$h_{ij} \leq \max\{h_{ik}, h_{jk}\} \quad \forall i, j, k \in \{1, 2, \cdots, n\},$$

where n is the number of records in a dataset and h_{ij} is the height of the internal node corresponding to the smallest cluster to which both record i and record j belong.

Figure 5.6 shows a dendrogram of the famous Iris dataset [79]. This dendrogram was created by the single linkage algorithm with the Euclidean distance. From the dendrogram we see that the single linkage algorithm produces two natural clusters for the Iris dataset.

More information about dendrograms can be found in [107], [106], [231], [149], [13], [244], [216], and [110]. Gordon [106] discussed the ultrametric conditions for dendrograms. Sibson [231] presented a mathematical representation of a dendrogram. Algorithms for plotting dendrograms are discussed in [216] and [110].

5.7 Partitional Clustering Algorithms

A partitional clustering algorithm is a clustering algorithm that divides a dataset into a single partition. Partitional clustering algorithms can be further classified into two categories: hard clustering algorithms and fuzzy clustering algorithms. In hard clustering, each record belongs to one and only one cluster. In fuzzy clustering, a record can belong to two or more clusters with probabilities.

Suppose a dataset with n records is clustered into k clusters by a partitional clustering algorithm. The clustering result of the partitional clustering algorithm can be represented by a $k \times n$ matrix U defined as

$$
U = \begin{pmatrix} u_{11} & u_{12} & \cdots & u_{1n} \\ u_{21} & u_{22} & \cdots & u_{2n} \\ \vdots & \vdots & \ddots & \vdots \\ u_{k1} & u_{k2} & \cdots & u_{kn} \end{pmatrix}. \tag{5.14}
$$

The matrix U produced by a hard clustering algorithm has the following properties:

$$
u_{ji} = 0 \text{ or } 1, \quad 1 \le j \le k, \, 1 \le i \le n, \tag{5.15a}
$$

$$
\sum_{j=1}^{k} u_{ji} = 1, \quad 1 \le i \le n, \tag{5.15b}
$$

$$
\sum_{i=1}^{n} u_{ji} > 0, \quad 1 \le j \le k. \tag{5.15c}
$$

The matrix U produced by a fuzzy clustering algorithm has the following properties:

$$
0 \le u_{ji} \le 1, \quad 1 \le j \le k, \, 1 \le i \le n, \tag{5.16a}
$$

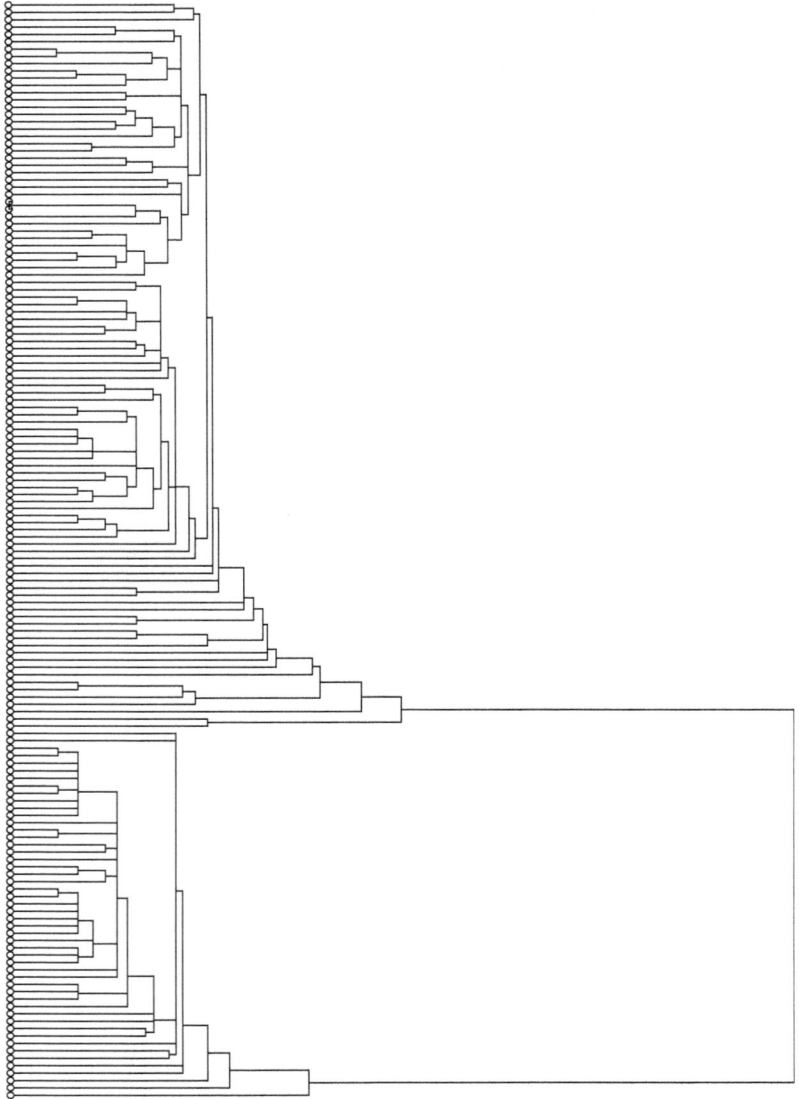

FIGURE 5.6: The dendrogram of the Iris dataset.

$$\sum_{j=1}^{k} u_{ji} = 1, \quad 1 \le i \le n, \qquad (5.16b)$$

$$\sum_{i=1}^{n} u_{ji} > 0, \quad 1 \le j \le k. \qquad (5.16c)$$

In this section, we present a survey of several partitional clustering algorithms. For a more comprehensive list of partitional clustering algorithms, readers are referred to [90] and [260].

5.7.1 Center-Based Clustering Algorithms

Center-based clustering algorithms [240, 272] are clustering algorithms that use a center to represent a cluster. Center-based clustering algorithms have two important properties [272]:

(a) They have a clearly defined objective function;

(b) They have a low runtime cost.

The standard k-means algorithm is a center-based clustering algorithm and is also one of the most popular and simple clustering algorithms. Although the k-means algorithm was first published in 1955 [145], about 50 years ago, it is still widely used today.

Given a dataset $X = \{\mathbf{x}_1, \mathbf{x}_2, \cdots, \mathbf{x}_n\}$ with n records. The k-means algorithm tries to divide the dataset into k disjoint clusters $C_1, C_2, \cdots, C_k$ by minimizing the following objective function

$$E = \sum_{i=1}^{k} \sum_{\mathbf{x} \in C_i} D(\mathbf{x}, \mu(C_i)),$$

where $D(\cdot, \cdot)$ is a distance function and $\mu(C_i)$ is the center of the cluster C_i and is usually defined as

$$\mu(C_i) = \frac{1}{|C_i|} \sum_{\mathbf{x} \in C_i} \mathbf{x}.$$

The standard k-means algorithm minimizes the objective function using an iterative process [25, 207, 229].

The standard k-means algorithm has several variations [90]. For example, the continuous k-means algorithm [76], the compare-means algorithm [207], the sort-means algorithm [207], the k-means algorithm based on kd-tree [206], and the trimmed k-means algorithm [48] are variations of the standard k-means algorithm.

Other center-based clustering algorithms include the k-modes algorithm [135, 41], the k-probabilities algorithm [255], the k-prototypes algorithm [136], the x-means algorithm [206], the k-harmonic means algorithm [271], the mean-shift algorithm [43, 44, 45, 85], and the maximum-entropy clustering (MEC) algorithm [218].

5.7.2   Search-based Clustering Algorithms

Many data clustering algorithms are formulated as some optimization problems [66, 197], which are complicate and have many local optimal solutions. Most of the clustering algorithms will stop when they find a locally optimal partition of the dataset. That is, most of the clustering algorithms may not be able to find the globally optimal partition of the dataset. For example, the fuzzy k-means algorithm [229] is convergent but may stop at a local minimum of the optimization problem.

Search-based clustering algorithms are developed to deal with the problem mentioned above. A search-based clustering algorithm aims at finding a globally optimal partition of a dataset by exploring the solution space of the underlying optimization problem. For example, clustering algorithms based on genetic algorithms [128] and tabu search [103] are search-based clustering algorithms.

Al-Sultan and Fedjki[9] proposed a clustering algorithm based on a tabu search technique. Ng and Wong [197] improved the fuzzy k-means algorithm using a tabu search algorithm. Other search-based clustering algorithms include the J-means algorithm [185], the genetic k-means algorithm [159], the global k-means algorithm [167], the genetic k-modes algorithm [93], and the SARS algorithm [132].

5.7.3   Graph-Based Clustering Algorithms

Clustering algorithms based on graph have also been proposed. A graph is a collection of vertices and edges. In graph-based clustering, a vertex represents a data point or record and a edge between a pair of vertices represents the similarity between the two records represented by the pair of vertices [260]. A cluster usually corresponds to a highly connected subgraph [123].

Several graph-based clustering algorithms have been proposed and developed. The chameleon [155] algorithm is a graph-based clustering algorithm that uses a sparse graph to represent a dataset. The CACTUS algorithm [99] is another graph-based clustering algorithm that uses a graph, called the similarity graph, to represent the inter-attribute and intra-attribute summaries. The ROCK algorithm [116] is an agglomerative hierarchical clustering algorithm that uses a graph connectivity to calculate the similarities between data points.

Gibson and coauthors [102] proposed a clustering algorithm based on hypergraphs and dynamical systems. Foggia and coauthors [81] proposed a graph-based clustering algorithm that is able to find clusters of any size and shape and does not require specifying the number of clusters. Foggia and coauthors [82] also compared the performance of several graph-based clustering algorithms.

Most of the graph-based clustering algorithms mentioned above use graphs as data structures and do not use graph analysis. Spectral clustering algorithms, which are also graph-based clustering algorithms, first construct a similarity graph and then use graph Laplacian matrices and standard linear algebra methods to divide a dataset into a number of clusters. von Luxburg [246] presented a tutorial on spectral clustering. Filippone and coauthors [78] also presented a survey of spectral clustering. Interested readers are referred to these two papers and the book by Ding and Zha [61].

5.7.4   Grid-Based Clustering Algorithms

Grid-based clustering algorithms are very efficient for clustering very large datasets since these algorithms perform clustering on the grid cells rather than the individual data points. A typical grid-based clustering algorithm consists of the following basic steps [111]:

(a) Construct a grid structure by dividing the data space into a finite number of cells;

(b) Calculate the density for each cell;

(c) Sort the cells based on their densities;

(d) Identify cluster centers;

(e) Traverse neighbor cells.

The STING (STatistical INformation Grid-based) algorithm [250] is a grid-based clustering algorithm proposed for clustering spatial datasets. STING was designed for clustering low-dimensional data and can not be scalable for clustering high-dimensional data. Keim and Hinneburg [157] proposed a grid-based clustering algorithm, OptiGrid, for clustering high-dimensional data. Schikuta and Erhart [225] proposed a BANG-clustering algorithm, which is also a grid-based clustering algorithm. Nagesh and coauthors [192] proposed a clustering algorithm based on adaptive grids. Other grid-based clustering algorithms include GRIDCLUS [224], GDILC [275], and WaveCluster [230].

More recently, Qiu and coauthors [209] also proposed a grid-based clustering algorithm that is capable of dealing with high dimensional datasets. Park and Lee [203] proposed a grid-based subspace clustering algorithm to cluster data streams. Lin and coauthors [168] proposed a grid-based clustering algorithm that is less influenced by the size of the grid cells than many other grid-based clustering algorithms.

5.7.5   Density-Based Clustering Algorithms

Density-based clustering algorithms are a kind of clustering algorithms that are capable of finding arbitrarily shaped clusters. In density-based clustering, a cluster is defined as a dense region surrounded by low-density regions. Usually, density-based clustering algorithms do not require specifying the number of clusters since these algorithms can automatically detect clusters and the number of clusters [70]. One drawback of most density-based clustering algorithm is that it is hard to determine certain parameters, such as the density threshold.

Popular density-based clustering algorithms include DBSCAN [71] and its variations and extensions such as GDBSCAN [222], PDBSCAN [263], DBCluC [270]. BRIDGE [54] is a hybrid clustering algorithm that is based on the k-means algorithm and DBSCAN. Other density-based clustering algorithms include DENCLUE [157] and CUBN [248].

5.7.6   Model-Based Clustering Algorithms

Mode-based clustering algorithms are clustering algorithms developed based on probability models. The term model usually refers to the type of constraints and geometric properties of the covariance matrices [177]. In model-based clustering, the data are viewed as samples coming from a mixture of probability distributions, each of which represents a cluster.

In model-based clustering, there are two approaches to formulate a model for the composite of clusters: the classification likelihood approach and the mixture likelihood approach [38, 84]. In the classification likelihood approach, the following objective function

$$\mathcal{L}_C(\Theta_1, \Theta_2, \cdots, \Theta_k; \gamma_1, \gamma_2, \cdots, \gamma_n | X) = \prod_{i=1}^{n} f_{\gamma_i}(\mathbf{x}_i | \Theta_{\gamma_i})$$

is maximized, where $\gamma_i = j$ if record $\mathbf{x}_i$ belongs to the jth component or cluster, Θ_j $(j = 1, 2, \cdots, k)$ are parameters, and $X = \{\mathbf{x}_1, \mathbf{x}_2, \cdots, \mathbf{x}_n\}$ is a dataset.

In the mixture likelihood approach, the following objective function

$$\mathcal{L}_M(\Theta_1, \Theta_2, \cdots, \Theta_k; \tau_1, \tau_2, \cdots, \tau_k | X) = \prod_{i=1}^{n} \sum_{j=1}^{k} \tau_j f_j(\mathbf{x}_i | \Theta_j)$$

is maximized, where $\tau_j \geq 0$ is the probability that a record belongs to the jth component and

$$\sum_{j=1}^{k} \tau_j = 1.$$

Some classical and powerful model-based clustering algorithms are based on Gaussian mixture models [14]. Celeux and Govaert [38] presented sixteen model-based clustering algorithm based on different constraints on the Gaussian mixture model. These algorithms use the EM algorithm [57, 180] to estimate the parameters.

A survey of data clustering based on a probabilistic and inferential framework can be found in [27]. Some early work on model-based clustering can be found in [24], [55], [68], [227], and [256]. Other model-based clustering algorithms are discussed in [90, Chapter 14].

5.7.7 Subspace Clustering Algorithms

Almost all conventional clustering algorithms do not work well for high dimensional datasets due to the following two problems associated with high dimensional data. First, the distance between any two data points in a high dimensional space becomes almost the same [20]. Second, clusters of high dimensional data are embedded in the subspaces of the data space and different clusters may exist in different subspaces [7]. Subspace clustering algorithms are clustering algorithms that are capable of finding clusters embedded in subspaces of the original data space.

Most subspace clustering algorithms can be classified into two major categories [204]: top-down algorithms and bottom-up algorithms. In top-down subspace clustering, a conventional clustering is performed and then the subspace of each cluster is evaluated. In bottom-up subspace clustering, dense regions in low dimensional spaces are identified and then these dense regions are combined to form clusters.

Examples of top-down subspace clustering algorithms include PART [35], PROCLUS [4], ORCLUS [5], FINDIT [257], and δ-cluster [264]. Examples of bottom-up subspace clustering algorithms include CLIQUE [7], ENCLUS [42], MAFIA [104], CLTree [169], DOC [208], and CBF [39].

There are also some subspace clustering algorithms that do not fit into the aforementioned categories. For example, the FSC (Fuzzy Subspace Clustering) algorithm [92, 91] is a subspace clustering, which is very similar to the k-means algorithm. The FSC algorithm uses a weight to represent the importance of a dimension or attribute to a cluster and incorporates the weights into the optimization problem.

Other subspace clustering algorithms include SUBCAD [96], the MSSC (Mean Shift for Subspace Clustering) algorithm [89, 90], and the grid-based subspace clustering algorithm [203]. Recent work on subspace clustering is presented in [58], [95], [152], [158], [191], and [205].

5.7.8 Neural Network-Based Clustering Algorithms

Neural network-based clustering algorithms are related to the concept of competitive learning [86, 112, 113, 220]. There are two types of competitive learning paradigms: hard competitive learning and soft competitive learning. Hard competitive learning is also known as winner-take-all or crisp competitive learning [15, 16]. In hard competitive learning, only a particular winning neuron that matches best with the given input pattern is allowed to learn. In soft competitive learning, all neurons have the opportunity to learn based on the input pattern. Hence soft competitive learning is also known as winner-take-most competitive learning [15, 16].

Example of neural network-based clustering algorithms include PART [34, 35], which is also a subspace clustering algorithm. The PARTCAT algorithm [97] is based on PART but was developed for clustering categorical data. Several other neural network-based clustering algorithms are presented and discussed in [260].

5.7.9   Fuzzy Clustering Algorithms

Most of the clustering algorithms presented in the previous several subsections are hard clustering algorithms, which require that each record belongs to one and only one cluster. Since [269] introduced the concept of fuzzy sets, fuzzy set theory has been applied to the area of data clustering [18, 221]. In fuzzy clustering, a record is allowed to belong to two or more clusters with probabilities.

Examples of fuzzy clustering algorithms include the fuzzy k-means algorithm [21, 22], the fuzzy k-modes algorithm [137], and the c-means algorithm [66, 67, 87, 124, 125, 142]. For more information about fuzzy clustering, readers are referred to the book by Höppner and and coauthors [130] and the survey papers [33] and [63].

5.8   Cluster Validity

Cluster validity is a collection of quantitative and qualitative measures that are used to evaluate and assess the results of clustering algorithms [145, 146]. Cluster validity indices can be defined based on three fundamental criteria: internal criteria, relative criteria, and external criteria [118, 119, 146, 241]. Both internal and external criteria are related to statistical testing.

In the external criteria approach, the results of a clustering algorithm are evaluated based on a prespecified structure imposed on the underlying dataset. Usually, external criteria require using the Monte Carlo simulation to do the evaluation [119]. Hence cluster validity based on external criteria is computationally expensive.

In the internal criteria approach, the results of a clustering algorithm are evaluated based on only quantities and features inherited from the underlying dataset. Cluster validity based on internal criteria can be used to assess results of hierarchical clustering algorithms as well as partitional clustering algorithms.

In the relative criteria approach, the results of a clustering algorithm are evaluated with other clustering results, which are produced by different clustering algorithms or the same algorithm but with different parameters. For example, a relative criterion is used to compare the results produced the

k-means algorithm with different parameter k (the number of clusters) in order to find the best clustering of the dataset.

5.9 Clustering Applications

Data clustering has been applied to many fields. According to [226], data clustering has been applied to the following five major groups:

(a) **Biology and zoology**. Clustering algorithms have been used to group animals and plants and develop taxonomies. In fact, data clustering algorithms were first developed in this field in which clustering is known as taxonomy analysis.

(b) **Medicine and psychiatry**. Clustering algorithms have been used to group diseases, including mental and physical diseases.

(c) **Sociology, criminology, anthropology, and archeology**. In fields in this group, data clustering algorithms have been used to group organizations, criminals, crimes, and cultures.

(d) **Geology, geography, and remote sensing**. In fields in this group, clustering algorithms have been used to group rock samples, sediments, cities, and land-use patterns.

(e) **Information retrieval, pattern recognition, market research, and economics**. Clustering algorithms have been used to analyze images, documents, industries, consumers, products, and markets.

The above list shows the applications of data clustering about 30 years ago. Nowadays, data clustering has a very broad applications. For example, data clustering has been applied to areas such as computational intelligence, machine learning, electrical engineering, genetics, and insurance [72, 260].

5.10 Literature on Data Clustering

Since the k-means algorithm was first published approximately 50 years ago [145], thousands of research papers and numerous books on data clustering have been published. In this section, we highlight some survey papers related to data clustering. For a comprehensive list of books, journals, and conference proceedings featuring research on data clustering, readers are directed to [90] and [94]. Additionally, many clustering algorithms are extensively discussed in the handbook of cluster analysis [126].

The following is a list of survey papers related to data clustering since 1971:

1971 A review of classification [46]

1982 A Survey of the Literature of Cluster Analysis [226]

1983 A survey of recent advances in hierarchical clustering algorithms [189]

1984 Counting dendrograms: A survey [190]

1987 A review of hierarchical classification [106]

1988 Recent trends in hierarchic document clustering: A critical review [254]

1993 A survey of fuzzy clustering [265]

1999 Data clustering: A review [144]

1999 A survey of fuzzy clustering algorithms for pattern recognition. I [15]

1999 A survey of fuzzy clustering algorithms for pattern recognition. II [16]

2000 Statistical pattern recognition: A review [147]

2004 Cluster analysis for gene expression data: A survey [151]

2004 Subspace clustering for high dimensional data: A review [204]

2005 Mining data streams: a review [88]

2005 Survey of Clustering Algorithms [261]

2008 A survey of kernel and spectral methods for clustering [78]

2009 Clustering high-dimensional data: A survey on subspace clustering, pattern-based clustering, and correlation clustering [158]

2009 A survey of Clustering Algorithms [217]

2010 Clustering Algorithms in Biomedical Research: A Review [262]

2010 A Survey of Clustering Algorithms for Graph Data [3]

2011 Spectral methods for graph clustering – A survey [195]

2012 A Survey on Enhanced Subspace Clustering [233]

2013 Data stream clustering: A survey [232]

2013 Functional data clustering: a survey [143]

2014 A Survey of Clustering Algorithms for Big Data: Taxonomy and Empirical Analysis [77]

2014 A survey on data stream clustering and classification [198]

2014 Model-based clustering of high-dimensional data: A review [31]

2014 A survey on nature inspired metaheuristic algorithms for partitional clustering [193]

2015 A Comprehensive Survey of Clustering Algorithms [259]

2015 A Survey of Multiobjective Evolutionary Clustering [187]

2015 Time-series clustering – A decade review [6]

2015 Combinatorial clustering: Literature review, methods, examples [165]

2016 A Survey on Feature Weighting Based K-Means Algorithms [56]

2016 A survey on soft subspace clustering [59]

2016 Automatic clustering using nature-inspired metaheuristics: A survey [154]

2016 Performance Analysis of Various Fuzzy Clustering Algorithms: A Review [108]

2016 Model-Based Clustering [179]

2017 Dominant-set clustering: A review [219]

2017 A review of clustering techniques and developments [223]

2017 Subspace multi-clustering: a review [131]

2017 A comprehensive survey of traditional, merge-split and evolutionary approaches proposed for determination of cluster number [120]

2018 A review of sparsity-based clustering methods [199]

2018 Multi-view clustering: A survey [266]

2018 Systematic Review of Clustering High-Dimensional and Large Datasets [202]

2018 A Survey of Clustering With Deep Learning: From the Perspective of Network Architecture [183]

2018 Triclustering Algorithms for Three-Dimensional Data Analysis: A Comprehensive Survey [127]

2018 Cluster ensembles: A survey of approaches with recent extensions and applications [30]

2019 To cluster, or not to cluster: An analysis of clusterability methods [1]

2019 Survey of State-of-the-Art Mixed Data Clustering Algorithms [8]

2020 Automatic clustering algorithms: a systematic review and bibliometric analysis of relevant literature [75]

2020 A survey of density based clustering algorithms [23]

2020 A survey on feature selection approaches for clustering [121]

2020 A survey on parallel clustering algorithms for Big Data [49]

2020 Clustering Categorical Data: A Survey [194]

2021 A Survey on Multiview Clustering [40]

2021 Scalable Clustering Algorithms for Big Data: A Review [175]

2022 A comprehensive survey of clustering algorithms: State-of-the-art machine learning applications, taxonomy, challenges, and future research prospects [74]

2022 Data clustering: application and trends [201]

2022 A survey of fuzzy clustering validity evaluation methods [247]

2022 Hierarchical clustering in astronomy [268]

2023 K-means clustering algorithms: A comprehensive review, variants analysis, and advances in the era of big data [140]

2023 Semi-supervised and un-supervised clustering: A review and experimental evaluation [239]

2023 Subspace Clustering in High-Dimensional Data Streams: A Systematic Literature Review [160]

2023 A review on semi-supervised clustering [32]

2024 A survey of genetic algorithms for clustering: Taxonomy and empirical analysis [215]

2024 Multi-modal data clustering using deep learning: A systematic review [213]

2024 Deep Clustering: A Comprehensive Survey [214]

2024 Deep image clustering: A survey [133]

2024 A survey on semi-supervised graph clustering [52]

2024 A comprehensive review of clustering techniques in artificial intelligence for knowledge discovery: Taxonomy, challenges, applications and future prospects [234]

2024 Three-way clustering: Foundations, survey and challenges [249]

2024 Survey of spectral clustering based on graph theory [62]

2024 Density peak clustering algorithms: A review on the decade 2014–2023 [251]

2024 Extended multivariate comparison of 68 cluster validity indices. A review [242]

2024 A review on declarative approaches for constrained clustering [53]

2024 Multiple clusterings: Recent advances and perspectives [267]

2024 Breaking down multi-view clustering: A comprehensive review of multi-view approaches for complex data structures [122]

2025 A survey of evidential clustering: Definitions, methods, and applications [274]

2025 Cluster validity indices for automatic clustering: A comprehensive review [141]

2025 Feature-weighted fuzzy clustering methods: An experimental review [105]

5.11 Summary

In this chapter, we introduced some fundamental concepts of data clustering, discussed various similarity and dissimilarity measures, and provided an overview of several clustering algorithms. While we did not delve deeply into most topics, we included references for readers seeking more detailed information. Notably, we compiled a list of survey papers related to data clustering, which serve as valuable resources for further exploration and understanding of the field.

6

Agglomerative Hierarchical Algorithms

A hierarchical clustering algorithm is a clustering algorithm that divides a dataset into a sequence of nested partitions. Hierarchical clustering algorithms can be divided into two categories: agglomerative hierarchical algorithms and divisive hierarchical algorithms [90].

An agglomerative hierarchical algorithm starts with every single record in a single cluster and repeats merging the closest pair of clusters according to some similarity or dissimilarity measure until all records are in one cluster. In contrast to an agglomerative hierarchical algorithm, a divisive hierarchical algorithm starts with all records in one cluster and repeats splitting a cluster into two smaller ones until all clusters contain only a single record.

In this chapter, we introduce the implementation of several agglomerative hierarchical algorithms. In particular, we focus on agglomerative hierarchical algorithms that use the Lance-Williams recurrence formula [161, 162].

6.1   Description of the Algorithm

Let $\{\mathbf{x}_0, \mathbf{x}_1, \cdots, \mathbf{x}_{n-1}\}$ be a dataset with n records. An agglomerative hierarchical algorithm starts with every single record as a cluster. Let $C_i = \{\mathbf{x}_i\}$, $i = 0, 2, \cdots, n-1$ be the n clusters at the beginning. Cluster $C_i (0 \leq i \leq n-1)$ contains the record $\mathbf{x}_i$. At each step, two clusters that have the minimum distance are merged to form a new cluster. An agglomerative hierarchical algorithm continues merging clusters until only one cluster is left.

For convenience, we assume that at step 1, a new cluster C_n is formed by merging two clusters in the initial set of clusters $\mathcal{F}_0 = \{C_0, C_1, \cdots, C_{n-1}\}$. Then after step 1 and before step 2, we have a set of clusters $\mathcal{F}_1 = \tilde{\mathcal{F}}_0 \cup \{C_n\}$, where $\tilde{\mathcal{F}}_0$ is the set of unmerged clusters in $\mathcal{F}_0$. If C_0 and C_1 have the minimum distance among all pairs of clusters, for example, then $C_n = C_0 \cup C_1$ and $\tilde{\mathcal{F}}_0 = \{C_2, C_3, \cdots, C_{n-1}\} = \mathcal{F}_0 \setminus \{C_0, C_1\}$.

At step 2, a new cluster C_{n+1} is formed by merging two clusters in the set of clusters $\mathcal{F}_1$. Similarly, we let $\tilde{\mathcal{F}}_1$ be the set of unmerged clusters in $\mathcal{F}_1$. Then

DOI: 10.1201/9781003592648-6

after step 2 and before step 3, we have a set of clusters $\mathcal{F}_2 = \tilde{\mathcal{F}}_1 \cup \{C_{n+1}\}$. The algorithm continues this process until at step $n - 1$ when the last two clusters are merged to form the cluster C_{2n-2}. After step $n - 1$, we have $\mathcal{F}_{n-1} = \{C_{2n-2}\}$, which contains only one cluster. The algorithm stops after step $n - 1$.

In the above process, we have $|\mathcal{F}_0| = n$, $|\mathcal{F}_1| = n - 1, \cdots$, and $|\mathcal{F}_{n-1}| = 1$, where $|\cdot|$ denotes the number of elements in the set. At each step, the algorithm merges two clusters. To decide which two clusters to be merged, we need to calculate the distances between clusters. Lance and Williams [161] proposed a recurrence formula to compute the distance between an old cluster and a new cluster formed by two old clusters.

The Lance-Williams formula is defined as follows. Before step i ($1 \leq i < n - 1$), we have a set of clusters $\mathcal{F}_{i-1}$, which contains $n - i + 1$ clusters. Suppose cluster C_{i_1} and C_{i_2} have the smallest distance among all the pairs of clusters in $\mathcal{F}_{i-1}$. Then C_{i_1} and C_{i_2} will be merged to form the cluster C_{n+i-1}. The Lance-Williams formula computes the distance between an old cluster $C \in \tilde{\mathcal{F}}_{i-1} = \mathcal{F}_{i-1} \setminus \{C_{i_1}, C_{i_2}\}$ as

$$
\begin{aligned}
D(C, C_{n+i-1}) &= D(C, C_{i_1} \cup C_{i_2}) \\
&= \alpha_{i_1} D(C, C_{i_1}) + \alpha_{i_2} D(C, C_{i_2}) + \beta D(C_{i_1}, C_{i_2}) \\
&\quad + \gamma |D(C, C_{i_1}) - D(C, C_{i_2})|,
\end{aligned} \tag{6.1}
$$

where α_{i_1}, α_{i_2}, β, and γ are parameters. DuBien and Warde [64] investigated some properties of the Lance-Williams formula.

TABLE 6.1: Parameters for the Lance-Williams formula, where $\Sigma = |C| + |C_{i_1}| + |C_{i_2}|$.

Algorithm	α_{i_1}	α_{i_2}	β	γ
Single linkage	$\frac{1}{2}$	$\frac{1}{2}$	0	$-\frac{1}{2}$
Complete linkage	$\frac{1}{2}$	$\frac{1}{2}$	0	$\frac{1}{2}$
Group average	$\frac{|C_{i_1}|}{|C_{i_1}|+|C_{i_2}|}$	$\frac{|C_{i_2}|}{|C_{i_1}|+|C_{i_2}|}$	0	0
Weighted group average	$\frac{1}{2}$	$\frac{1}{2}$	0	0
Centroid	$\frac{|C_{i_1}|}{|C_{i_1}|+|C_{i_2}|}$	$\frac{|C_{i_2}|}{|C_{i_1}|+|C_{i_2}|}$	$-\frac{|C_{i_1}|\cdot|C_{i_2}|}{(|C_{i_1}|+|C_{i_2}|)^2}$	0
Median	$\frac{1}{2}$	$\frac{1}{2}$	0	$-\frac{1}{4}$
Ward's method	$\frac{|C|+|C_{i_1}|}{\Sigma}$	$\frac{|C|+|C_{i_2}|}{\Sigma}$	$-\frac{|C|}{\Sigma}$	0

Table 6.1 gives seven sets of parameters for the Lance-Williams formula defined in Equation (6.1). Each set of parameters results in an agglomerative hierarchical clustering algorithm. A more general recurrence formula was proposed by [148] and discussed in [90] and [107].

The first four algorithms (i.e., single linkage, complete linkage, group average, and weighted group average) are referred to as graph hierarchical methods [189]. The last three algorithms (i.e., centroid, median, and Ward's method) are referred to as geometric hierarchical methods. The last three algorithms requires squared Euclidean distance in the Lance-Williams formula. For geometric hierarchical algorithms, the centers of a cluster formed by merging two clusters can be calculated from the centers of the two merged clusters. In addition, the distance between two clusters can be calculated from the distance between centers of the two clusters. Table 6.2 shows the calculation of centers and distances of clusters for the geometric hierarchical algorithms.

TABLE 6.2: Centers of combined clusters and distances between two clusters for geometric hierarchical algorithms, where $\mu(\cdot)$ denotes a center of a cluster and $D_{euc}(\cdot, \cdot)$ is the Euclidean distance.

Algorithm	$\mu(C_1 \cup C_2)$	$D(C_1, C_2)$
Centroid	$\frac{|C_1|\mu(C_1)+|C_2|\mu(C_2)}{|C_1|+|C_2|}$	$D_{euc}(\mu(C_1), \mu(C_2))^2$
Median	$\frac{\mu(C_1)+\mu(C_2)}{2}$	$D_{euc}(\mu(C_1), \mu(C_2))^2$
Ward's	$\frac{|C_1|\mu(C_1)+|C_2|\mu(C_2)}{|C_1|+|C_2|}$	$\frac{|C_1|\cdot|C_2|}{|C_1|+|C_2|}D_{euc}(\mu(C_1), \mu(C_2))^2$

A hierarchical clustering algorithm is said to be monotonic if at each step we have

$$D(C, C_1 \cup C_2) \geq D(C_1, C_2), \tag{6.2}$$

where C_1 and C_2 are the two clusters to be merged and C is an unmerged cluster. The single linkage algorithm and the complete linkage algorithm are monotonic [153]. However, other agglomerative hierarchical algorithms might violate the monotonic inequality [182].

6.2 Implementation

Agglomerative clustering algorithms have been implemented in the Python packages `scikit-learn` and `SciPy`. For illustration purpose, we implement some agglomerative algorithms in this section.

The input to an agglomerative clustering algorithm is a condensed distance matrix, which can be obtained by calling the `pdist` function from the SciPy library. The condensed distance matrix calculated by the `pdist` function is a one-dimensional array by combining all the rows from the upper triangle of the distance matrix. To convert the one-dimensional array to the square form distance matrix, we can use the `squareform` function.

The indices of the square form distance matrix and those of the corresponding one-dimensional array are calculated as follows. Let M be the square form distance matrix and V be the corresponding one-dimensional array that stores the upper triangle of M. Let m be the number of rows of M. Then the length of V is $\dfrac{n(n-1)}{2}$. For $0 \le i < j < m$, the (i,j)th entry of M is saved to the entry of V with the following index:

$$mi - \frac{(i+1)(i+2)}{2} + j. \tag{6.3}$$

The hth entry of V corresponds to the $(i_h, j_h)$the entry of M, where i_h and j_h are determined as follows:

$$i_h = \left\lfloor \frac{2m - 1 - \sqrt{(2m-1)^2 - 8h}}{2} \right\rfloor, \tag{6.4a}$$

$$j_h = \frac{(i+1)(i+2)}{2} - mi_h + h. \tag{6.4b}$$

The indices given in Equation (6.4) are obtained by letting

$$h = mi - \frac{(i+1)(i+2)}{2} + j$$

and solving the following inequalities:

$$i + 1 \le \frac{(i+1)(i+2)}{2} - mi + h = j \le m - 1.$$

The following Python function implements the conversion of the indices between M and V:

```python
def getInd(ind, m):
    ind = np.ascontiguousarray(ind)
    if len(ind) == 2: # convert (i,j) -> h
        return int(m*ind[0] + ind[1] - (ind[0]+1)*(ind[0]+2)/2)
    else: # convert h -> (i, j)
        i = int(np.floor((2*m-1-np.sqrt((2*m-1)**2 - 8*ind[0]))/2))
        j = int((i+1)*(i+2)/2 -m*i+ind[0])
        return (i, j)
```

If the input to the above function is a tuple (or a list) of coordinates of M, the output is the index of V. If the input is an index of V, the output is a tuple of the coordinates of M.

6.2.1   The Single Linkage Algorithm

According to the Lance-Williams recurrence formula, the single linkage algorithm calculates the distance between a new cluster C_{n+i-1} formed at step i and an old cluster C as

$$
\begin{aligned}
&D(C, C_{n+i-1}) \\
={}& D(C, C_{i_1} \cup C_{i_2}) \\
={}& \frac{1}{2}D(C, C_{i_1}) + \frac{1}{2}D(C, C_{i_2}) - \frac{1}{2}|D(C, C_{i_1}) - D(C, C_{i_2})| \\
={}& \min\{D(C, C_{i_1}), D(C, C_{i_2})\},
\end{aligned}
\tag{6.5}
$$

where C_{i_1} and C_{i_2} are the two clusters merged at step i.

Following the C++ code in [94], we can implement the single linkage algorithm in Python as follows:

```python
def single(dm):
    dm2 = squareform(dm)
    n = dm2.shape[0]
    unmergedClusters = list(range(0,n))
    clusterSize = [1] * n
    res = np.zeros((n-1, 4))
    for s in range(n-1):
        m = len(unmergedClusters)
        R, C = np.triu_indices(m, k=1)
        dist = dm2[R, C]
        indMin = np.argmin(dist)
        dMin = dist[indMin]
        ij = getInd(indMin, m)
        s1 = unmergedClusters[ij[0]]
        s2 = unmergedClusters[ij[1]]
        size = clusterSize[s1] + clusterSize[s2]
        res[s,:] = [s1, s2, dMin, size]
        unmergedClusters.remove(s1)
        unmergedClusters.remove(s2)
        unmergedClusters.append(n+s)
        clusterSize.append(size)
        # update distance matrix
        m -= 1
        tmp = np.zeros((m, m))
        ind = list(range(m+1))
        ind.remove(ij[0])
        ind.remove(ij[1])
        tmp[0:(m-1), 0:(m-1)] = dm2[np.ix_(ind, ind)]
        tmp[0:(m-1), m-1] = np.minimum(dm2[ind, ij[0]], dm2
            [ind, ij[1]])
        tmp[m-1, 0:(m-1)] = tmp[0:(m-1), m-1]
        dm2 = tmp
    return res
```

The logic of the above code is similar to that given in [94]. Since Python is a scripting language, we try to avoid loops as much as possible in the above implementation. Instead, we use the vectorization feature of NumPy to speed up the calculation. One drawback of this approach is that we use more memory.

6.2.2 The Complete Linkage Algorithm

The complete linkage algorithm calculates the distance between a new cluster C_{n+i-1} formed at step i and an old cluster C as

$$
\begin{aligned}
& D(C, C_{n+i-1}) \\
= & D(C, C_{i_1} \cup C_{i_2}) \\
= & \frac{1}{2} D(C, C_{i_1}) + \frac{1}{2} D(C, C_{i_2}) + \frac{1}{2} |D(C, C_{i_1}) - D(C, C_{i_2})| \\
= & \max\{D(C, C_{i_1}), D(C, C_{i_2})\},
\end{aligned}
\tag{6.6}
$$

where C_{i_1} and C_{i_2} are the two clusters merged at step i.

The implementation of the complete linkage algorithm is given below:

```
def complete(dm):
    dm2 = squareform(dm)
    n = dm2.shape[0]
    unmergedClusters = list(range(0,n))
    clusterSize = [1] * n
    res = np.zeros((n-1, 4))
    for s in range(n-1):
        m = len(unmergedClusters)
        R, C = np.triu_indices(m, k=1)
        dist = dm2[R, C]
        indMin = np.argmin(dist)
        dMin = dist[indMin]
        ij = getInd(indMin, m)
        s1 = unmergedClusters[ij[0]]
        s2 = unmergedClusters[ij[1]]
        size = clusterSize[s1] + clusterSize[s2]
        res[s,:] = [s1, s2, dMin, size]
        unmergedClusters.remove(s1)
        unmergedClusters.remove(s2)
        unmergedClusters.append(n+s)
        clusterSize.append(size)
        # update distance matrix
        m -= 1
        tmp = np.zeros((m, m))
        ind = list(range(m+1))
        ind.remove(ij[0])
        ind.remove(ij[1])
        tmp[0:(m-1), 0:(m-1)] = dm2[np.ix_(ind, ind)]
        tmp[0:(m-1), m-1] = np.maximum(dm2[ind, ij[0]], dm2
            [ind, ij[1]])
```

```
30        tmp[m-1, 0:(m-1)] = tmp[0:(m-1), m-1]
31        dm2 = tmp
32    return res
```

The implementation of the complete linkage algorithm is exactly the same as the single linkage algorithm except that we use the `maximum` function to update the distance matrix.

6.2.3   The Group Average Algorithm

The group average algorithm calculates the distance between a new cluster C_{n+i-1} formed at step i and an old cluster C as

$$
\begin{aligned}
D(C, C_{n+i-1}) \\
= \ & D(C, C_{i_1} \cup C_{i_2}) \\
= \ & \frac{|C_{i_1}|}{|C_{i_1}| + |C_{i_2}|} D(C, C_{i_1}) + \frac{|C_{i_2}|}{|C_{i_1}| + |C_{i_2}|} D(C, C_{i_2}) \\
= \ & \frac{|C_{i_1}| \cdot D(C, C_{i_1}) + |C_{i_2}| \cdot D(C, C_{i_2})}{|C_{i_1}| + |C_{i_2}|},
\end{aligned}
\tag{6.7}
$$

where C_{i_1} and C_{i_2} are the two clusters merged at step i and $|\cdot|$ denotes the number of elements in the underlying set.

The implementation of the group average algorithm is given below:

```
1  def average(dm):
2      dm2 = squareform(dm)
3      n = dm2.shape[0]
4      unmergedClusters = list(range(0,n))
5      clusterSize = [1] * n
6      res = np.zeros((n-1, 4))
7      for s in range(n-1):
8          m = len(unmergedClusters)
9          R, C = np.triu_indices(m, k=1)
10         dist = dm2[R, C]
11         indMin = np.argmin(dist)
12         dMin = dist[indMin]
13         ij = getInd(indMin, m)
14         s1 = unmergedClusters[ij[0]]
15         s2 = unmergedClusters[ij[1]]
16         size = clusterSize[s1] + clusterSize[s2]
17         res[s,:] = [s1, s2, dMin, size]
18         unmergedClusters.remove(s1)
19         unmergedClusters.remove(s2)
20         unmergedClusters.append(n+s)
21         clusterSize.append(size)
22         # update distance matrix
23         m -= 1
```

```
24      tmp = np.zeros((m, m))
25      ind = list(range(m+1))
26      ind.remove(ij[0])
27      ind.remove(ij[1])
28      tmp[0:(m-1), 0:(m-1)] = dm2[np.ix_(ind, ind)]
29      tmp[0:(m-1), m-1] = (clusterSize[s1]*dm2[ind, ij
            [0]] + clusterSize[s2]*dm2[ind, ij[1]])/size
30      tmp[m-1, 0:(m-1)] = tmp[0:(m-1), m-1]
31      dm2 = tmp
32  return res
```

6.2.4   The Weighted Group Average Algorithm

The weighted group average algorithm is also called the weighted pair group method using arithmetic average [146]. According to the Lance-Williams recurrence formula, the weighted group average algorithm calculates the distance between a new cluster C_{n+i-1} formed at step i and an old cluster C as

$$
\begin{aligned}
D(C, C_{n+i-1}) &= D(C, C_{i_1} \cup C_{i_2}) \\
&= \frac{1}{2}D(C, C_{i_1}) + \frac{1}{2}D(C, C_{i_2}),
\end{aligned}
\tag{6.8}
$$

where C_{i_1} and C_{i_2} are the two clusters merged at step i.

The implementation of the weighted group average algorithm is given below:

```
1  def weighted(dm):
2      dm2 = squareform(dm)
3      n = dm2.shape[0]
4      unmergedClusters = list(range(0,n))
5      clusterSize = [1] * n
6      res = np.zeros((n-1, 4))
7      for s in range(n-1):
8          m = len(unmergedClusters)
9          R, C = np.triu_indices(m, k=1)
10         dist = dm2[R, C]
11         indMin = np.argmin(dist)
12         dMin = dist[indMin]
13         ij = getInd(indMin, m)
14         s1 = unmergedClusters[ij[0]]
15         s2 = unmergedClusters[ij[1]]
16         size = clusterSize[s1] + clusterSize[s2]
17         res[s,:] = [s1, s2, dMin, size]
18         unmergedClusters.remove(s1)
19         unmergedClusters.remove(s2)
20         unmergedClusters.append(n+s)
21         clusterSize.append(size)
```

```
22        # update distance matrix
23        m -= 1
24        tmp = np.zeros((m, m))
25        ind = list(range(m+1))
26        ind.remove(ij[0])
27        ind.remove(ij[1])
28        tmp[0:(m-1), 0:(m-1)] = dm2[np.ix_(ind, ind)]
29        tmp[0:(m-1), m-1] = (dm2[ind, ij[0]] + dm2[ind, ij
          [1]])/2
30        tmp[m-1, 0:(m-1)] = tmp[0:(m-1), m-1]
31        dm2 = tmp
32    return res
```

6.2.5   The Centroid Algorithm

The centroid algorithm is also called the unweighted pair group method using centroids [146]. According to the Lance-Williams recurrence formula, the centroid algorithm calculates the distance between a new cluster C_{n+i-1} formed at step i and an old cluster C as

$$
\begin{aligned}
& D(C, C_{n+i-1}) \\
=\ & D(C, C_{i_1} \cup C_{i_2}) \\
=\ & \frac{|C_{i_1}|}{|C_{i_1}| + |C_{i_2}|} D(C, C_{i_1}) + \frac{|C_{i_2}|}{|C_{i_1}| + |C_{i_2}|} D(C, C_{i_2}) \\
& - \frac{|C_{i_1}| \cdot |C_{i_2}|}{(|C_{i_1}| + |C_{i_2}|)^2} D(C_{i_1}, C_{i_2}),
\end{aligned}
\tag{6.9}
$$

where C_{i_1} and C_{i_2} are the two clusters merged at step i, and $D(\cdot, \cdot)$ is the squared Euclidean distance defined in Equation (5.7).

The implementation of the centroid algorithm is given below:

```
1  def centroid(dm):
2      dm2 = squareform(np.square(dm))
3      n = dm2.shape[0]
4      unmergedClusters = list(range(0,n))
5      clusterSize = [1] * n
6      res = np.zeros((n-1, 4))
7      for s in range(n-1):
8          m = len(unmergedClusters)
9          R, C = np.triu_indices(m, k=1)
10         dist = dm2[R, C]
11         indMin = np.argmin(dist)
12         dMin = np.sqrt(dist[indMin])
13         ij = getInd(indMin, m)
14         s1 = unmergedClusters[ij[0]]
15         s2 = unmergedClusters[ij[1]]
16         size = clusterSize[s1] + clusterSize[s2]
```

```
17      res[s,:] = [s1, s2, dMin, size]
18      unmergedClusters.remove(s1)
19      unmergedClusters.remove(s2)
20      unmergedClusters.append(n+s)
21      clusterSize.append(size)
22      # update distance matrix
23      m -= 1
24      tmp = np.zeros((m, m))
25      ind = list(range(m+1))
26      ind.remove(ij[0])
27      ind.remove(ij[1])
28      tmp[0:(m-1), 0:(m-1)] = dm2[np.ix_(ind, ind)]
29      tmp[0:(m-1), m-1] = (clusterSize[s1]*dm2[ind, ij
            [0]] + clusterSize[s2]*dm2[ind, ij[1]] -
            clusterSize[s1]*clusterSize[s2]*dm2[ij[0], ij
            [1]]/size)/size
30      tmp[m-1, 0:(m-1)] = tmp[0:(m-1), m-1]
31      dm2 = tmp
32  return res
```

The implementation of the centroid algorithm is different from those of the previous agglomerative algorithms in several places. First, the distance is changed to squared Euclidean distance. However, the distance saved to the dendrogram is still the Euclidean distance.

6.2.6   The Median Algorithm

The median algorithm has another name, which is called the weighted pair group method using centroids [146]. According to the Lance-Williams recurrence formula, the median algorithm calculates the distance between a new cluster C_{n+i-1} formed at step i and an old cluster C as

$$
\begin{aligned}
& D(C, C_{n+i-1}) \\
= & D(C, C_{i_1} \cup C_{i_2}) \\
= & \frac{1}{2} D(C, C_{i_1}) + \frac{1}{2} D(C, C_{i_2}) - \frac{1}{4} D(C_{i_1}, C_{i_2}), \quad (6.10)
\end{aligned}
$$

where C_{i_1} and C_{i_2} are the two clusters merged at step i, and $D(\cdot, \cdot)$ is the squared Euclidean distance defined in Equation (5.7).

Similar to the centroid algorithm, the median algorithm is implemented as follows:

```
1  def median(dm):
2      dm2 = squareform(np.square(dm))
3      n = dm2.shape[0]
4      unmergedClusters = list(range(0,n))
5      clusterSize = [1] * n
6      res = np.zeros((n-1, 4))
```

```
7     for s in range(n-1):
8         m = len(unmergedClusters)
9         R, C = np.triu_indices(m, k=1)
10        dist = dm2[R, C]
11        indMin = np.argmin(dist)
12        dMin = np.sqrt(dist[indMin])
13        ij = getInd(indMin, m)
14        s1 = unmergedClusters[ij[0]]
15        s2 = unmergedClusters[ij[1]]
16        size = clusterSize[s1] + clusterSize[s2]
17        res[s,:] = [s1, s2, dMin, size]
18        unmergedClusters.remove(s1)
19        unmergedClusters.remove(s2)
20        unmergedClusters.append(n+s)
21        clusterSize.append(size)
22        # update distance matrix
23        m -= 1
24        tmp = np.zeros((m, m))
25        ind = list(range(m+1))
26        ind.remove(ij[0])
27        ind.remove(ij[1])
28        tmp[0:(m-1), 0:(m-1)] = dm2[np.ix_(ind, ind)]
29        tmp[0:(m-1), m-1] = dm2[ind, ij[0]]/2 + dm2[ind, ij
             [1]]/2 - dm2[ij[0], ij[1]]/4
30        tmp[m-1, 0:(m-1)] = tmp[0:(m-1), m-1]
31        dm2 = tmp
32    return res
```

6.2.7   Ward's Algorithm

The Ward's algorithm was proposed by [252] and [253]. This algorithm aims to minimize the loss of information associated with each merging. Hence the Ward's algorithm is also referred to as the "minimum variance" method.

According to the Lance-Williams recurrence formula, the Ward's algorithm algorithm calculates the distance between a new cluster C_{n+i-1} formed at step i and an old cluster C as

$$
\begin{aligned}
&D(C, C_{n+i-1}) \\
&= D(C, C_{i_1} \cup C_{i_2}) \\
&= \frac{|C| + |C_{i_1}|}{|C| + |C_{i_1}| + |C_{i_2}|} D(C, C_{i_1}) + \frac{|C| + |C_{i_2}|}{|C| + |C_{i_1}| + |C_{i_2}|} D(C, C_{i_2}) \\
&\quad - \frac{|C|}{|C| + |C_{i_1}| + |C_{i_2}|} D(C_{i_1}, C_{i_2}), \quad\quad\quad (6.11)
\end{aligned}
$$

where C_{i_1} and C_{i_2} are the two clusters merged at step i, and $D(\cdot, \cdot)$ is the squared Euclidean distance defined in Equation (5.7).

The Ward's algorithm is implemented as follows:

```python
def ward(dm):
    dm2 = squareform(np.square(dm))
    n = dm2.shape[0]
    unmergedClusters = list(range(0,n))
    clusterSize = [1] * n
    res = np.zeros((n-1, 4))
    for s in range(n-1):
        m = len(unmergedClusters)
        R, C = np.triu_indices(m, k=1)
        dist = dm2[R, C]
        indMin = np.argmin(dist)
        dMin = np.sqrt(dist[indMin])
        ij = getInd(indMin, m)
        s1 = unmergedClusters[ij[0]]
        s2 = unmergedClusters[ij[1]]
        size = clusterSize[s1] + clusterSize[s2]
        res[s,:] = [s1, s2, dMin, size]
        unmergedClusters.remove(s1)
        unmergedClusters.remove(s2)
        unmergedClusters.append(n+s)
        clusterSize.append(size)
        # update distance matrix
        m -= 1
        tmp = np.zeros((m, m))
        ind = list(range(m+1))
        ind.remove(ij[0])
        ind.remove(ij[1])
        ind2 = unmergedClusters[0:-1]
        tmp[0:(m-1), 0:(m-1)] = dm2[np.ix_(ind, ind)]
        size2 = np.array([clusterSize[i] for i in ind2])
        tmp[0:(m-1), m-1] = np.divide( np.multiply(size2+
            clusterSize[s1], dm2[ind, ij[0]]) + np.multiply
            (size2+clusterSize[s2], dm2[ind, ij[1]]) -
            size2 * dm2[ij[0], ij[1]], size2+clusterSize[s1
            ]+clusterSize[s2])
        tmp[m-1, 0:(m-1)] = tmp[0:(m-1), m-1]
        dm2 = tmp
    return res
```

In the above implementation, we use the NumPy functions `divide` and `multiply` to perform element-wise operations on arrays.

6.3   Examples

In this section, we present some examples of applying the agglomerative hierarchical clustering algorithms implemented in the previous section. In particular, we will compare the results produced by our code with those produced by the agglomerative algorithms provided by the SciPy library.

To test our implementations and perform the comparison, we use the Iris dataset from the UCI machine learning repository. First, let us load the necessary libraries by executing the following code:

```
import numpy as np
import pandas as pd
import matplotlib.pyplot as plt
import scipy.cluster.hierarchy as hierarchy
from scipy.spatial.distance import pdist, squareform
from ucimlrepo import fetch_ucirepo
```

Then we load the Iris dataset and calculate the distance matrix as follows:

```
iris = fetch_ucirepo(id=53)
X = iris.data.features
dm = pdist(X)
```

The distance matrix `dm` is a condensed distance matrix.

After executing all the code given in the previous section, we will have the Python functions `single`, `complete`, `group`, `weighted`, `centroid`, `median`, and `ward`. Those functions correspond to different agglomerative clustering algorithms.

Now we can test the single linkage clustering algorithm with the Iris dataset as follows:

```
Z1 = single(dm)
Z1b = hierarchy.linkage(dm, 'single')

fig, axes = plt.subplots(2, 1, figsize=(8, 8))
dn1 = hierarchy.dendrogram(Z1, ax=axes[0], no_labels=True,
    color_threshold=0, above_threshold_color="black")
dn1b = hierarchy.dendrogram(Z1b, ax=axes[1], no_labels=True
    , color_threshold=0, above_threshold_color="black")
fig.savefig("single.pdf", bbox_inches='tight')
```

The first dendrogram `Z1` is produced by our code. The second dendrogram `Z1b` is produced by the SciPy library. Figure 6.1 shows these two dendrograms. From the figure, we see that the two dendrograms are almost the same. However, they have some slight differences, which are caused by the merging of clusters that have same distances to other clusters.

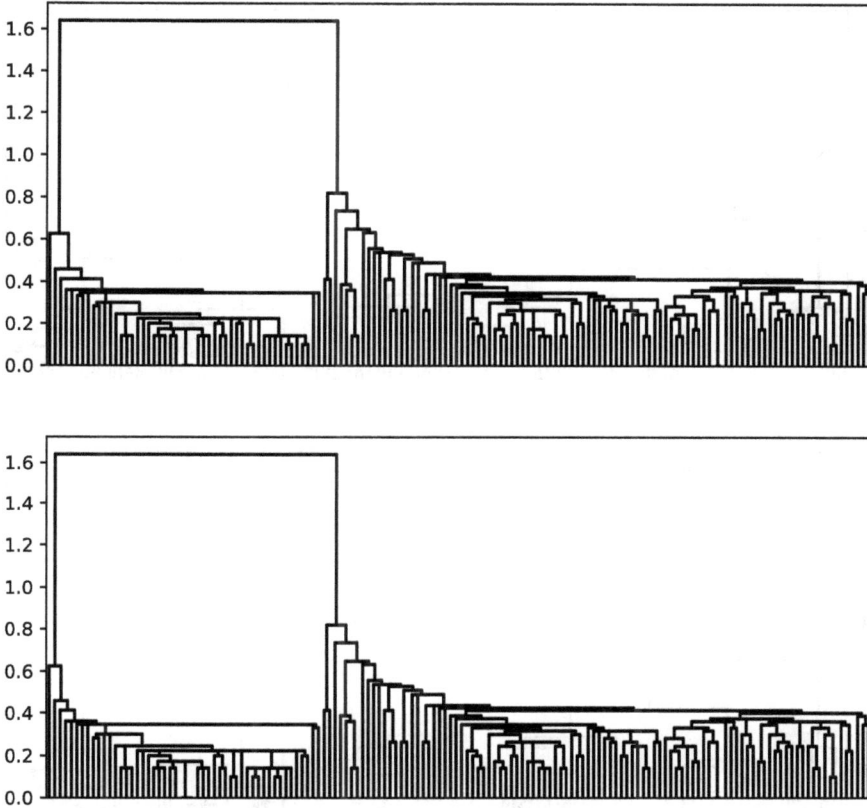

FIGURE 6.1: Dendrograms produced by the single linkage clustering algorithm. The top dengrogram is produced by code from this book. The bottom one is produced by the SciPy library.

To test the complete linkage algorithm, we use the following code:

```
Z1 = complete(dm)
Z1b = hierarchy.linkage(dm, 'complete')

fig, axes = plt.subplots(2, 1, figsize=(8, 8))
dn1 = hierarchy.dendrogram(Z1, ax=axes[0], no_labels=True,
    color_threshold=0, above_threshold_color="black")
dn1b = hierarchy.dendrogram(Z1b, ax=axes[1], no_labels=True
    , color_threshold=0, above_threshold_color="black")
fig.savefig("complete.pdf", bbox_inches='tight')
```

The dendrograms produced by the complete linkage algorithm are shown in Figure 6.2. From the figure, we see that the dendrogram produced by our code matches that produced by the SciPy library. It is hard to notice the difference.

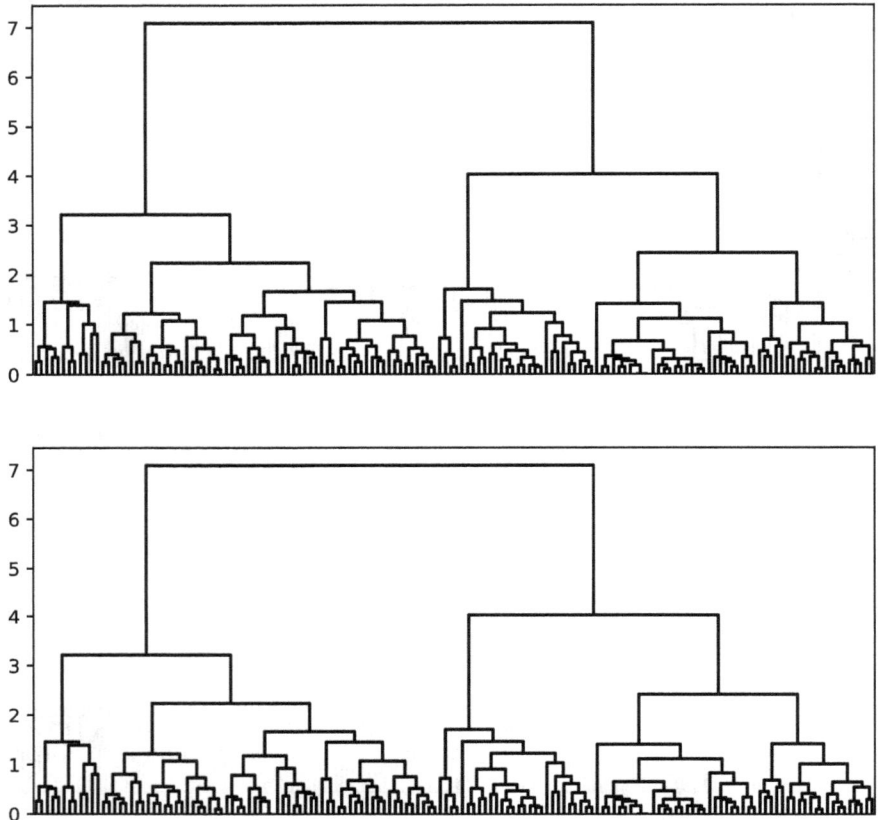

FIGURE 6.2: Dendrograms produced by the complete linkage clustering algorithm. The top dengrogram is produced by code from this book. The bottom one is produced by the SciPy library.

To cluster the Iris dataset using the group average algorithm, we can use the following code:

```
1  Z1 = average(dm)
2  Z1b = hierarchy.linkage(dm, 'average')
3
4  fig, axes = plt.subplots(2, 1, figsize=(8, 8))
5  dn1 = hierarchy.dendrogram(Z1, ax=axes[0], no_labels=True,
       color_threshold=0, above_threshold_color="black")
6  dn1b = hierarchy.dendrogram(Z1b, ax=axes[1], no_labels=True
       , color_threshold=0, above_threshold_color="black")
7  fig.savefig("average.pdf", bbox_inches='tight')
```

The resulting dendrograms obtained by the group average clustering algorithm are shown in Figure 6.3.

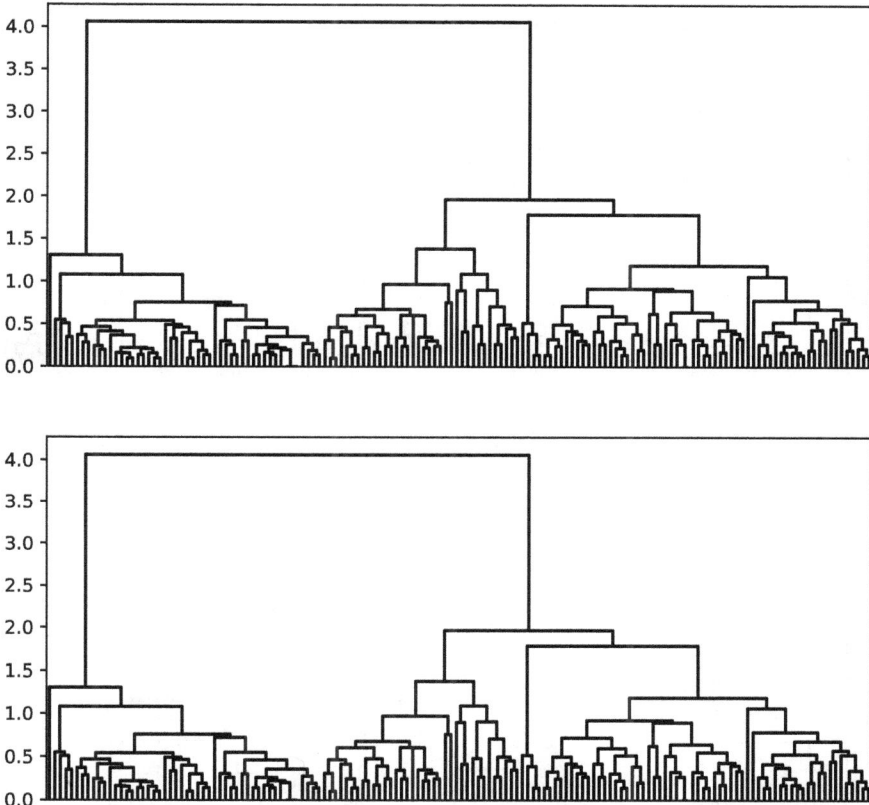

FIGURE 6.3: Dendrograms produced by the group average clustering algorithm. The top dengrogram is produced by code from this book. The bottom one is produced by the SciPy library.

To cluster the Iris dataset using the weighted group average algorithm, we can use the following code:

```
Z1 = weighted(dm)
Z1b = hierarchy.linkage(dm, 'weighted')

fig, axes = plt.subplots(2, 1, figsize=(8, 8))
dn1 = hierarchy.dendrogram(Z1, ax=axes[0], no_labels=True,
    color_threshold=0, above_threshold_color="black")
dn1b = hierarchy.dendrogram(Z1b, ax=axes[1], no_labels=True
    , color_threshold=0, above_threshold_color="black")
fig.savefig("weighted.pdf", bbox_inches='tight')
```

Figure 6.4 shows the dendrograms produced by the weighted group average algorithm.

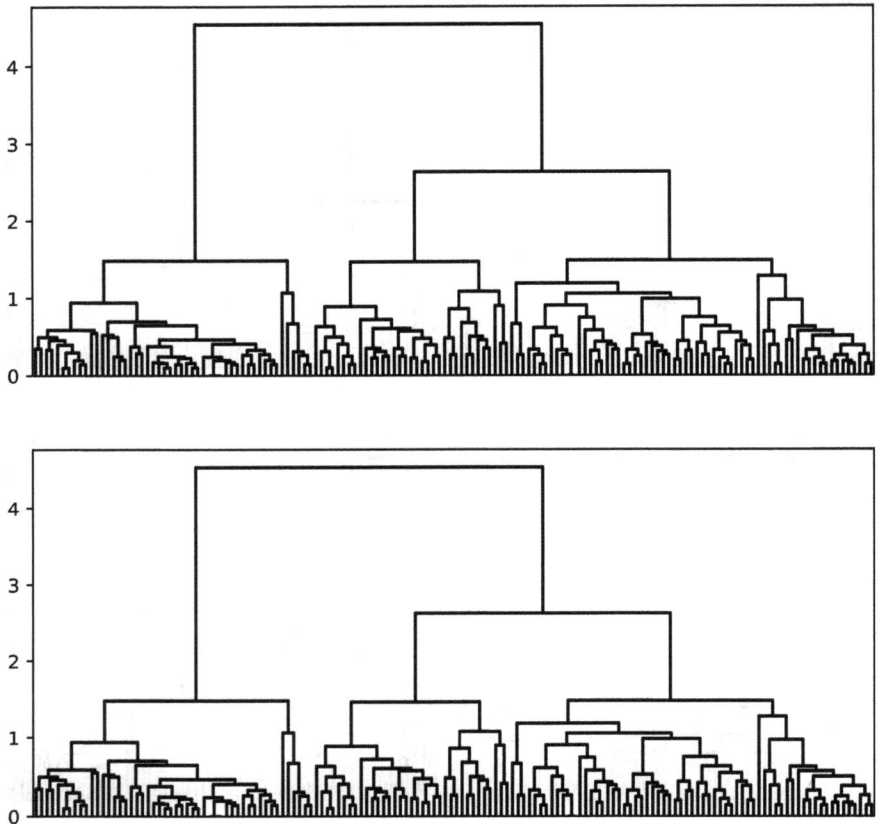

FIGURE 6.4: Dendrograms produced by the weighted group average clustering algorithm. The top dengrogram is produced by code from this book. The bottom one is produced by the SciPy library.

The centroid algorithm, the median algorithm, and the Ward's algorithm use squared Euclidean distances to calculate the distances between merged clusters and other clusters. We will just test one of these algorithm and let the reader to test the other algorithms. Let us test the Ward's algorithm as follows:

```
Z1 = ward(dm)
Z1b = hierarchy.linkage(dm, 'ward')

fig, axes = plt.subplots(2, 1, figsize=(8, 8))
dn1 = hierarchy.dendrogram(Z1, ax=axes[0], no_labels=True,
```

```
     color_threshold=0, above_threshold_color="black")
6  dn1b = hierarchy.dendrogram(Z1b, ax=axes[1], no_labels=True
       , color_threshold=0, above_threshold_color="black")
7  fig.savefig("ward.pdf", bbox_inches='tight')
```

The resulting dendrograms produced by Ward's algorithm are shown in Figure 6.5. Again, the dendrogram produced by our code matches that produced by the SciPy library.

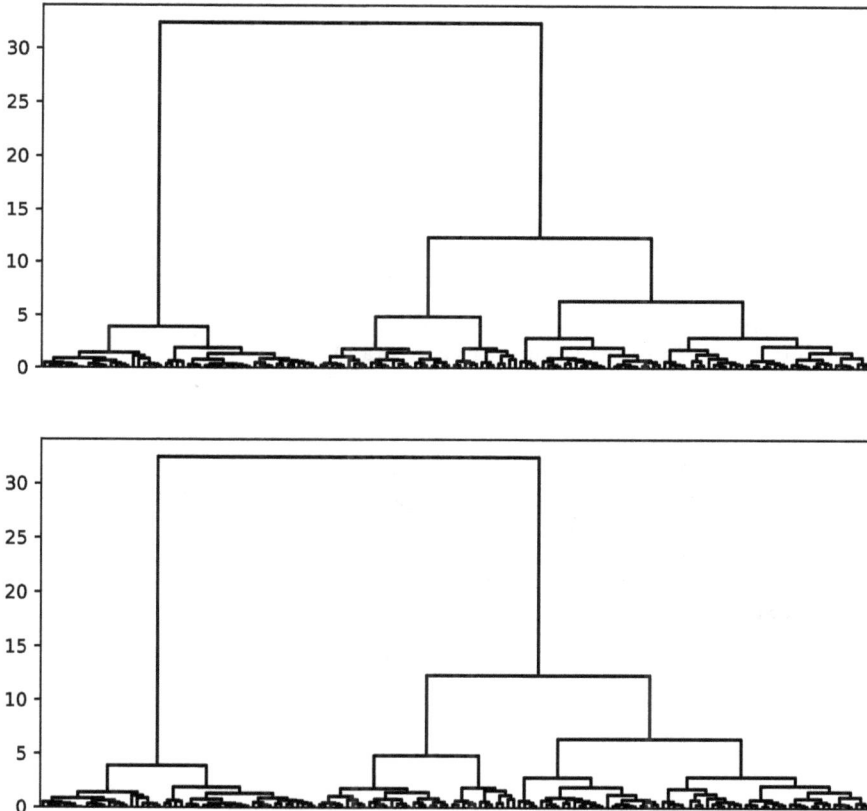

FIGURE 6.5: Dendrograms produced by the Ward's clustering algorithm. The top dengrogram is produced by code from this book. The bottom one is produced by the SciPy library.

Hierarchical clusters can be converted to flat clusters by cutting the dendrogram. For example, the function **fcluster** from the SciPy library can be used to form flat clusters from a dendrogram. The following Python code illustrates how to create flat clusters from a dendrogram:

```
1  y = iris.data.targets
2  ms1 = hierarchy.fcluster(Z1, t=6.5, criterion='distance')
3  cm1 = createCM(y, ms1)
4  print(cm1)
```

The function **createCM** is used to create a confusion matrix from the observed labels and the predicted labels. The code of this function is given in Listing 6.1. For the Ward's algorithm, the confusion matrix obtained by the above code is given below:

```
1                     1    2    3
2  Iris-setosa       50    0    0
3  Iris-versicolor    0    1   49
4  Iris-virginica     0   35   15
```

The confusion matrix is implemented as a Pandas data frame. The confusion matrix shows that 16 records of the Iris dataset are misclassified by the Ward's algorithm.

```
1  def createCM(y, yhat):
2      y = np.ascontiguousarray(y)
3      yhat = np.ascontiguousarray(yhat)
4      labelr = np.unique(y)
5      labelc = np.unique(yhat)
6      nrow = len(labelr)
7      ncol = len(labelc)
8      cm = pd.DataFrame(np.zeros((nrow, ncol), dtype=np.int32
           ), index=labelr, columns=labelc)
9      for i in range(len(y)):
10         cm.loc[y[i], yhat[i]] += 1
11     return cm
```

Listing 6.1: Python code used to create a confusion matrix from observed labels and predicted lables.

6.4 Summary

In this chapter, we presented the implementation of several agglomerative hierarchical clustering algorithms based on the Lance-Williams framework. We also presented examples of applying these agglomerative hierarchical algorithms to a synthetic dataset and the Iris dataset. In our examples, we only tried the Euclidean distance to calculate the pair-wise distances between records and tried to get a partition of maximum three clusters from the hierarchical clustering tree. We encourage users to try other distances and apply these algorithms to cluster other datasets.

7

A Divisive Hierarchical Clustering Algorithm

Hierarchical clustering algorithms can be agglomerative or divisive. In the previous chapter, we implemented several agglomerative hierarchical clustering algorithms. In this chapter, we implement a divisive hierarchical clustering algorithm, DIANA (DIVisive ANAlysis), which was described in [156, Chapter 6].

7.1 Description of the Algorithm

DIANA (DIVisive ANAlysis) is a divisive hierarchical clustering algorithm based on the idea of [173]. Given a dataset consisting of n records, there are $2^{n-1} - 1$ ways to divide the dataset into two nonempty groups. The DIANA algorithm does not consider all these divisions.

Precisely, let $X = \{\mathbf{x}_1, \mathbf{x}_1, \cdots, \mathbf{x}_{n-1}\}$ be a dataset with n records. At the beginning, all the n records are in one cluster. In the first step, the algorithm divides the dataset into two groups using some iterative process. To do this, the algorithm first finds the record that has the greatest average distance to the rest records. The average distance between record $\mathbf{x}_i$ to the rest is calculated as

$$D_i = \frac{1}{n-1} \sum_{j=0, j \neq i}^{n-1} D(\mathbf{x}_i, \mathbf{x}_j),$$

where $D(\cdot, \cdot)$ is some distance measure.

Suppose $D_1 = \max_{0 \leq i \leq n-1} D_i$, i.e., $\mathbf{x}_0$ has the greatest average distance to the rest records in the dataset. Then $\mathbf{x}_0$ is first spitted from the dataset. Then we have two groups now: $G_1 = \{\mathbf{x}_0\}$ and $G_2 = X \setminus G_1 = \{\mathbf{x}_1, \mathbf{x}_2, \cdots, \mathbf{x}_{n-1}\}$. Then the algorithm checks every record in G_2 to see if the record should be moved to G_1. To do this, the algorithm calculates the distance between $\mathbf{x}$ and G_1 and the distance between $\mathbf{x}$ and $G_2 \setminus \{\mathbf{x}\}$ for all $\mathbf{x} \in G_2$. The distance

DOI: 10.1201/9781003592648-7

between $\mathbf{x}$ and G_1 is calculated as

$$D_{G_1}(\mathbf{x}) = \frac{1}{|G_1|} \sum_{\mathbf{y} \in G_1} D(\mathbf{x}, \mathbf{y}), \quad \mathbf{x} \in G_2 \qquad (7.1)$$

where $|G_1|$ denotes the number of records in G_1. The distance between $\mathbf{x}$ and $G_2 \setminus \{\mathbf{x}\}$ is calculated as

$$D_{G_2}(\mathbf{x}) = \frac{1}{|G_2| - 1} \sum_{\mathbf{y} \in G_2} D(\mathbf{x}, \mathbf{y}). \quad \mathbf{x} \in G_2 \qquad (7.2)$$

If $D_{G_1}(\mathbf{x}) < D_{G_2}(\mathbf{x})$, then $\mathbf{x}$ is moved from G_2 to G_1. The algorithm continues checking all other records in G_2 until no records should be moved. At this stage, the dataset is divided into two clusters: G_1 and G_2.

In the second step, the algorithm first finds the cluster that has the largest diameter. The diameter of a cluster is defined as the maximum distance between any two records in the cluster. That is,

$$Diam(G) = \max_{\mathbf{x}, \mathbf{y} \in G} D(\mathbf{x}, \mathbf{y}). \qquad (7.3)$$

If $Diam(G_1) > Diam(G_2)$, then the algorithm applies the process described in the first step to divide cluster G_1 into two clusters: G_3 and G_4.

The algorithm repeats the above procedure until every cluster contains only one record. The algorithm can finish the process in $n - 1$ steps.

7.2   Implementation

To implement the divisive hierarchical clustering algorithm, we need to assign a unique identifier to each cluster in the hierarchical clustering tree and a level to these clusters. To do this, we follow the same approach we used to implement the agglomerative hierarchical clustering algorithms. That is, given a dataset $X = \{\mathbf{x}_0, \mathbf{x}_1, \cdots, \mathbf{x}_{n-1}\}$, the initial cluster that contains all the records has an identifier of $2n - 2$. We denote the first cluster by X_{2n-2}.

At the first step, X_{2n-2} is divided into two clusters. If all the two clusters have more than one record, then the two clusters are denoted as X_{2n-3} and X_{2n-4}. If one of the two clusters has only one record, then we denote the one-record cluster by X_i if the cluster contains $\mathbf{x}_i$. If the other cluster contains more than one record, then we denote the other cluster by X_{2n-3}. At the end, the clustering tree includes clusters $X_{2n-2}, X_{2n-3}, \cdots, X_{n-1}, \cdots$, and X_0. The clusters $X_0, X_1, \cdots, X_{n-1}$ are one-record clusters. All other clusters contain more than one record.

In Python, we implement the DIANA algorithm as follows:

```python
def diana(dm):
    dm2 = squareform(dm)
    n = dm2.shape[0]
    clusterDiameter = [0] * (2*n-1)
    clusterMember = [[]] * (2*n-1)
    clusterDiameter[2*n-2] = np.max(dm)
    clusterMember[2*n-2] = [i for i in range(n)]
    unsplitClusters = [2*n-2]
    res = np.zeros((n-1, 4))
    res[n-2, 2:4] = [clusterDiameter[2*n-2], n]
    for s in range(n-2,-1,-1):
        diam =[clusterDiameter[i] for i in unsplitClusters]
        indMax = np.argmax(diam)
        s0 = unsplitClusters[indMax]
        ms0 = clusterMember[s0]
        ms1, ms2 = split(dm2, ms0)
        if len(ms1) > 1:
            s1 = 2*n - 2 - clusterMember[::-1].index([])
            clusterMember[s1] = ms1
            clusterDiameter[s1] = np.max(dm2[np.ix_(ms1,
                ms1)])
            res[s1-n, 2:4] = [clusterDiameter[s1], len(ms1)
                ]
            unsplitClusters.append(s1)
        else:
            s1 = ms1[0]
        if len(ms2) > 1:
            s2 = 2*n - 2 - clusterMember[::-1].index([])
            clusterMember[s2] = ms2
            clusterDiameter[s2] = np.max(dm2[np.ix_(ms2,
                ms2)])
            res[s2-n, 2:4] = [clusterDiameter[s2], len(ms2)
                ]
            unsplitClusters.append(s2)
        else:
            s2 = ms2[0]
        res[s0-n, 0:2] = [s1, s2]
        unsplitClusters.remove(s0)
    return res
```

In the above code, we use the list `clusterDiameter` to store the diameters of all clusters. We use a list of lists to store cluster members. For a dataset with n records, there will be $2n - 1$ clusters, which are indexed from 0 to $2n - 2$. Clusters with indices from 0 to $n - 1$ are singleton clusters. The function used to split a cluster is defined in Listing 7.1.

```
1  def split(dm, ms):
2      dm2 = dm[np.ix_(ms, ms)]
3      dist = np.sum(dm2, axis=0)
4      indMax = np.argmax(dist)
5      ms1 = [ms[indMax]]
6      ms2 = ms.copy()
7      ms2.remove(ms[indMax])
8      bChanged = True
9      while bChanged:
10         bChanged = False
11         if len(ms2) == 1:
12             break
13         for s in ms2:
14             dist1 = np.sum(dm[s, ms1])/len(ms1)
15             dist2 = np.sum(dm[s, ms2])/(len(ms2)-1)
16             if dist1 < dist2:
17                 bChanged = True
18                 ms1.append(s)
19                 ms2.remove(s)
20                 break
21      return ms1, ms2
```

Listing 7.1: The function used to split a cluster into two.

The logic of the above Python code is similar to that of the C++ code given in [94].

7.3   Examples

In this section, we apply the DIANA algorithm implemented in the previous section to the Iris dataset.

First, we load the necessary Python libraries by running the following code:

```
1  import numpy as np
2  import pandas as pd
3  import matplotlib.pyplot as plt
4  import scipy.cluster.hierarchy as hierarchy
5  from scipy.spatial.distance import pdist, squareform
6  from ucimlrepo import fetch_ucirepo
```

Then we run all the code given in the previous section and the following code to apply the DIANA algorithm to the Iris dataset:

```
1  iris = fetch_ucirepo(id=53)
2  X = iris.data.features
3  dm = pdist(X)
4
5  Z = diana(dm)
```

```
6
7  fig, axes = plt.subplots(1, 1, figsize=(8, 4))
8  dn1 = hierarchy.dendrogram(Z, ax=axes, no_labels=True,
       color_threshold=0, above_threshold_color="black")
9  fig.savefig("diana.pdf", bbox_inches='tight')
```

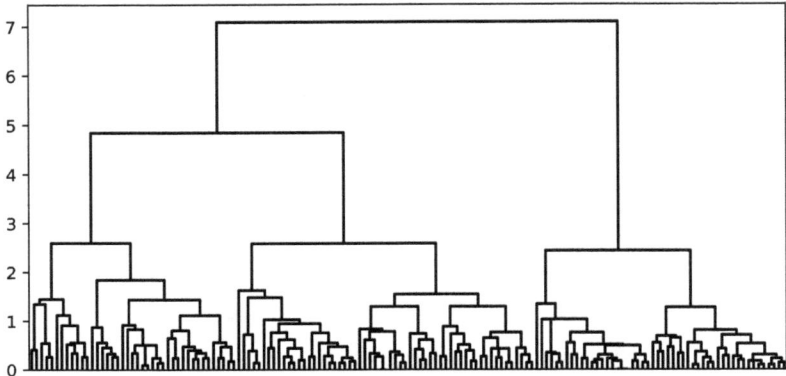

FIGURE 7.1: The dendrogram produced by applying the DIANA algorithm in Python to the Iris dataset.

The resulting dendrogram is shown in Figure 7.1. The dendrogram produced by the C++ code given in [94] is shown in Figure 7.2. The two dendrograms are the same except for the orientation and the vertical scale.

We can convert hierarchical clusters to flat clusters and then obtain the confusion matrix as follows:

```
1  y = iris.data.targets
2  ms1 = hierarchy.fcluster(Z, t=3.5, criterion='distance')
3  cm1 = createCM(y, ms1)
4  print(cm1)
```

The function **createCM** is defined in Listing 6.1. Executing the above block of code gives the following output:

```
1                    1     2    3
2  Iris-setosa        0     0   50
3  Iris-versicolor    5    45    0
4  Iris-virginica    36    14    0
```

The confusion matrix shows that 19 records are misclassified by the DIANA algorithm.

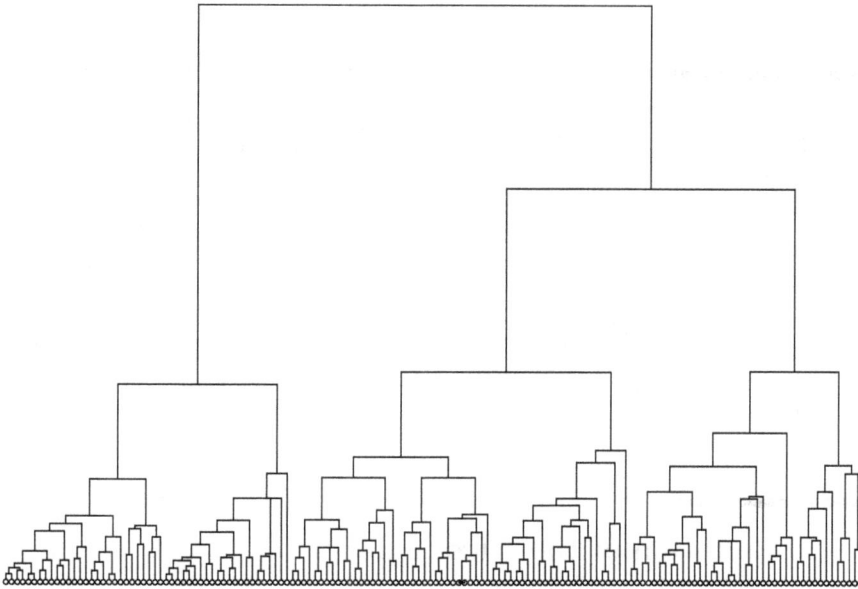

FIGURE 7.2: The dendrogram produced by applying the DIANA algorithm in C++ to the Iris dataset.

7.4   Summary

In this chapter, we implemented a divisive hierarchical clustering algorithm in Python. Examples to illustrate the DIANA algorithm were also presented.

The DIANA algorithm was based on the idea presented in [173]. However, the original idea is to split all available clusters. Kaufman and Rousseeuw [156] modified this original method by defining the diameter for a cluster and first splitting the cluster with the largest diameter. The modified method produces a monotonic hierarchical clustering. One drawback of this modified method is that it is sensitive to outliers.

8

The k-means Algorithm

The k-means algorithm is the most popular and the simplest partitional clustering algorithm [145]. The k-means algorithm has many variations (see Section 5.7). In this chapter, we implement the standard k-means algorithm.

8.1 Description of the Algorithm

Let $X = \{\mathbf{x}_0, \mathbf{x}_1, \cdots, \mathbf{x}_{n-1}\}$ be a numeric dataset containing n records and k be an integer in $\{1, 2, \cdots n\}$. The k-means algorithm tries to divide the dataset into k clusters $C_0, C_1, \cdots,$ and C_{k-1} by minimizing the following objective function

$$E = \sum_{i=1}^{k-1} \sum_{\mathbf{x} \in C_i} D(\mathbf{x}, \boldsymbol{\mu}_i), \qquad (8.1)$$

where $D(\cdot, \cdot)$ is a distance measure and $\mu(C_i)$ is the mean of cluster C_i, i.e.,

$$\boldsymbol{\mu}_i = \frac{1}{|C_i|} \sum_{\mathbf{x} \in C_i} \mathbf{x}.$$

Let γ_i be the cluster membership of record $\mathbf{x}_i$ for $i = 0, 1, \cdots, n-1$. That is, $\gamma_i = j$ if $\mathbf{x}_i$ belongs to cluster C_j. Then Equation (8.1) can be rewirtten as

$$E = \sum_{i=0}^{n-1} D(\mathbf{x}_i, \boldsymbol{\mu}_{\gamma_i}). \qquad (8.2)$$

To minimize the objective function, the k-means algorithm employs an iterative process. At the beginning, the k-means algorithm select k random records from the dataset X as initial cluster centers.

Suppose $\boldsymbol{\mu}_0^{(0)}$, $\boldsymbol{\mu}_1^{(0)}$, $\cdots$, and $\boldsymbol{\mu}_{k-1}^{(0)}$ are the initial cluster centers. Based on these cluster centers, the k-means algorithm updates the cluster memberships $\gamma_0^{(0)}, \gamma_1^{(0)}, \cdots, \gamma_{n-1}^{(0)}$ as follows:

$$\gamma_i^{(0)} = \underset{0 \le j \le k-1}{\operatorname{argmin}} D(\mathbf{x}_i, \boldsymbol{\mu}_j^{(0)}), \qquad (8.3)$$

DOI: 10.1201/9781003592648-8

where argmin is the argument that minimizes the distance. That is, $\gamma_i^{(0)}$ is set to the index of the cluster to which $\mathbf{x}_i$ has the smallest distance.

Based on the cluster memberships $\gamma_0^{(0)}, \gamma_1^{(0)}, \cdots, \gamma_{n-1}^{(0)}$, the k-means algorithm updates the cluster centers as follows:

$$\boldsymbol{\mu}_j^{(1)} = \frac{1}{|\{i : \gamma_i^{(0)} = j\}|} \sum_{i=0, \gamma_i^{(0)}=j}^{n-1} \mathbf{x}_i, \quad j = 0, 1, \cdots, k - 1. \qquad (8.4)$$

Then the k-means algorithm repeats updating the cluster memberships based on Equation (8.3) and updating the cluster centers based on Equation (8.4) until some condition is satisfied. For example, the k-means algorithm stops when the cluster memberships do not change any more.

8.2 Implementation

To implement the standard k-means algorithm described in the previous section, we follow the logic of the C++ code given in [94].

In Python, we implement the k-means algorithm as follows:

```
def kmeans(X, k=3, maxit=100):
    X = np.ascontiguousarray(X)
    n, d = X.shape
    ind = np.random.choice(n, k, replace=False)
    clusterCenters = X[ind,:]
    dm = np.zeros((n,k))
    for i in range(k):
        dm[:,i] = np.sum(np.square(X-clusterCenters[i,:]),
            axis=1)
    clusterMembership = np.argmin(dm, axis=1)
    numIter = 1
    while numIter < maxit:
        # update cluster centers
        for i in range(k):
            bInd = clusterMembership==i
            if np.any(bInd):
                clusterCenters[i,:] = np.mean(X[bInd], axis
                    =0)
            else:
                clusterCenters[i,:] = X[np.random.randint
                    (0, n),:]
        # update cluster membership
        clusterMembership_ = clusterMembership.copy()
        for i in range(k):
            dm[:,i] = np.sum(np.square(X-clusterCenters[i
                ,:]), axis=1)
```

```
23    clusterMembership = np.argmin(dm, axis=1)
24    nChanges = np.count_nonzero(clusterMembership -
          clusterMembership_)
25    numIter += 1
26    if nChanges == 0:
27        break;
28    objectiveValue = np.sum(dm[list(range(n)),
        clusterMembership]).item()
29    return clusterMembership, clusterCenters,
        objectiveValue, numIter
```

In the above code, we assume that the input data X is a two-dimensional array with numerical elements. At the beginning, we initialize k cluster centers by selecting k distinct records from the input dataset. Then we initialize the cluster memberships by assigning records to their nearest centers.

After the cluster centers and the cluster memberships are initialized, we use a while loop to repeatedly update the cluster centers and the cluster memberships until the cluster memberships no longer change. However, we impose a maximum number of iterations.

The k-means algorithm may produce empty clusters, especially when the number of clusters is relatively large. In our implementation, we prevent empty clusters by updating the centers of empty clusters to randomly selected data points. This mechanism is implemented in Lines 14–18 of the above code.

In the above implementation, we try to avoid loops as much as possible. Since the desired number of clusters is usually much smaller than the number of records, we loop through the number of clusters rather than the number of records.

Since the k-means algorithm can produce different results based on different initial cluster centers, we can run the algorithm multiple times to select the best result. We can implement the multiple run of the k-means algorithm as follows:

```
1  def kmeans2(X, k=3, numrun=10, maxit=100):
2      bestCM, bestCC, bestOV, bestIters = kmeans(X, k, maxit)
3      print([bestOV, bestIters])
4      for i in range(numrun-1):
5          cm, cc, ov, iters = kmeans(X, k, maxit)
6          print([ov, iters])
7          if ov < bestOV:
8              bestCM, bestCC, bestOV, bestIters = cm, cc, ov,
                  iters
9      return bestCM, bestCC, bestOV, bestIters
```

In the above code, we select the run with the minimum objective function value as the best run.

8.3 Examples

In this section, we show how to use the k-means algorithm implemented in the previous section. Before running the code given in the previous section, we load the necessary libraries by executing the following code:

```
import numpy as np
import pandas as pd
import matplotlib.pyplot as plt
from sklearn.datasets import make_blobs
from ucimlrepo import fetch_ucirepo
```

First, we apply the k-means algorithm to a synthetic dataset. The following piece of code illustrates how to generate a synthetic dataset and apply the k-means algorithm to the dataset:

```
centers = [[3, 3], [-3, -3], [3, -3]]
X, y = make_blobs(n_samples=300, centers=centers,
    cluster_std=1, random_state=1)

yhat, cc, ov, iters = kmeans(X, 3)
cm1 = createCM(y, yhat)
print([ov, iters])
print(cm1)

bcm, bcc, bov, biters = kmeans2(X)
cm2 = createCM(y, bcm)
print([bov, biters])
print(cm2)
```

Executing the above block of code gives the following output:

```
[580.3494342617068, 10]
        0    1    2
0   0    100   0
1   99    0    1
2   0     1   99
[580.3494342617068, 2]
[580.3494342617068, 4]
[580.3494342617068, 4]
[580.3494342617068, 4]
[580.3494342617068, 9]
[580.3494342617068, 9]
[580.3494342617068, 5]
[580.3494342617068, 5]
[580.3494342617068, 4]
[580.3494342617068, 7]
[580.3494342617068, 2]
```

```
17          0    1    2
18  0   100    0    0
19  1     0   99    1
20  2     1    0   99
```

From the output, we see that the k-means algorithm produced very good results on the synthetic dataset. Only two records were misclassified by the k-means algorithm. In all the runs, the k-means algorithm produced the same objective function value. However, the number of iterations were different for different runs.

Since the synthetic dataset is a two-dimensional dataset, we can visualize the clustering results by plotting the data points with different markers. We can produce such a plot as follows:

```
1  fig, ax = plt.subplots(1, 1, figsize=(6, 6))
2  markers = ["x", "o", "+"]
3  for i in range(3):
4      members = bcm == i
5      center = bcc[i,:]
6      ax.plot(X[members, 0], X[members, 1], markers[i], color
           ="black")
7      ax.plot(center[0], center[1], "^", markerfacecolor="
           white",
8          markeredgecolor="black", markersize=15)
9  fig.savefig("kmeans1.pdf", bbox_inches='tight')
```

After executing the above block of code, we see the plot shown in Figure 8.1. The cluster centers are denoted by triangles. Cluster memberships are indicated by different markers.

In the rest of this section, we apply the k-means algorithm to the Iris dataset. Let us apply the k-means algorithm to the Iris dataset 20 times:

```
1  iris = fetch_ucirepo(id=53)
2  X = iris.data.features
3  y = iris.data.targets
4
5  bcm, bcc, bov, biters = kmeans2(X, numrun=20)
6  cm1 = createCM(y, bcm)
7  print(cm1)
8  print([bov, biters])
```

Executing the above block of code produced the following output:

```
1  [78.94084142614602, 4]
2  [78.94084142614602, 3]
3  [78.94506582597731, 9]
4  [78.94084142614602, 6]
5  [78.94506582597731, 5]
6  [78.94506582597731, 10]
```

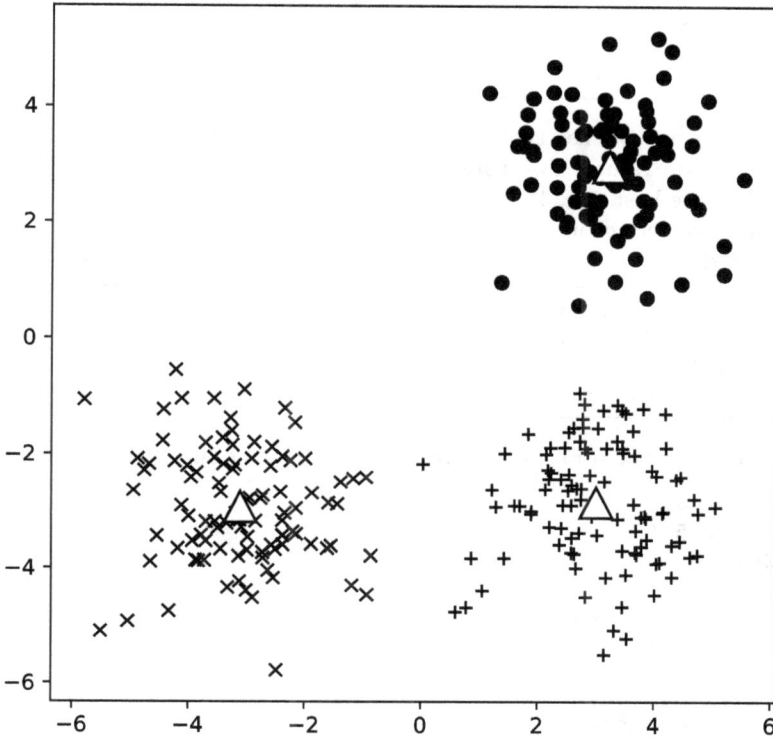

FIGURE 8.1: Clustering results produced by applying the k-means algorithm to the synthetic data.

```
 7 [78.94506582597731 , 10]
 8 [78.94506582597731 , 12]
 9 [78.94084142614602 , 4]
10 [78.94084142614602 , 7]
11 [78.94084142614602 , 6]
12 [143.45373548406212 , 6]
13 [142.85929166666665 , 4]
14 [78.94084142614602 , 7]
15 [78.94084142614602 , 5]
16 [78.94506582597731 , 6]
17 [78.94506582597731 , 16]
18 [78.94084142614602 , 3]
19 [78.94506582597731 , 5]
20 [78.94084142614602 , 7]
21                          0   1   2
22 Iris-setosa             0  50   0
```

```
23 Iris-versicolor    48    0    2
24 Iris-virginica     14    0   36
25 [78.94084142614602, 4]
```

From the output, we see that most of the runs produced relatively small objective function values. Two runs produced relatively large objective function values. For the best run, 16 records are misclassified. Records with the target Iris-setosa were grouped 100% correctly by the k-means algorithm.

To divide the Iris dataset into six clusters, we can run the following code:

```
1 bcm, bcc, bov, biters = kmeans2(X, k=6, numrun=20)
2 cm1 = createCM(y, bcm)
3 print(cm1)
4 print([bov, biters])
```

After executing the above code, we see the following output:

```
 1 [68.65823307743929, 15]
 2 [42.312521556681986, 8]
 3 [39.251830892636775, 7]
 4 [68.245945237742, 11]
 5 [47.62757749753634, 11]
 6 [48.43080641716744, 8]
 7 [39.185257692307694, 7]
 8 [45.42942304384424, 8]
 9 [48.091082025432456, 12]
10 [47.75236080174198, 5]
11 [47.66450931571816, 11]
12 [47.75415897978063, 16]
13 [45.47813516505636, 6]
14 [39.251830892636775, 6]
15 [42.312521556681986, 8]
16 [68.44723527042174, 11]
17 [38.957011157119865, 7]
18 [45.45010085470085, 8]
19 [42.36690446650125, 7]
20 [41.801410101010106, 6]
```

	0	1	2	3	4	5
Iris-setosa	0	0	27	23	0	0
Iris-versicolor	0	27	0	0	0	23
Iris-virginica	12	1	0	0	24	13

```
25 [38.957011157119865, 7]
```

Among the 20 runs of the k-means algorithm, the best run produced an objective function value of 38.957. Records with the target Iris-setosa were divided into two clusters. Records with the target Iris-versicolor were also divided into two clusters. Records with the target Iris-virginica were divided into three major clusters.

The k-means algorithm is implemented in the scikit-learn library. To apply the k-means algorithm from this library to the Iris dataset, we can run the following code:

```
from sklearn.cluster import KMeans
res = KMeans(n_clusters=6, random_state=0, n_init="auto").
    fit(X)
cm2 = createCM(y, res.labels_)
print(cm2)
print([res.inertia_, res.n_iter_])
```

After executing the above block of code, we see the following output:

```
                  0   1   2   3   4   5
Iris-setosa       0  27   0   0   0  23
Iris-versicolor  27   0   0   2  21   0
Iris-virginica    1   0  22  27   0   0
[42.31252155668197, 7]
```

The results are a little different from those produced by the k-means algorithm implemented in this chapter. To see the difference, we can calculate the confusion matrix between the results produced by the two versions of k-means:

```
print(createCM(bcm, res.labels_))
```

The resulting confusion matrix is

```
     0   1   2   3   4   5
0    0   0  12   0   0   0
1   28   0   0   0   0   0
2    0  27   0   0   0   0
3    0   0   0   0   0  23
4    0   0  10  14   0   0
5    0   0   0  15  21   0
```

From the confusion matrix, we see that four clusters are the same. Two clusters are different.

8.4 Summary

In this chapter, we introduced the standard k-means algorithm and its implementation. We also applied the k-means algorithm to a synthetic dataset and the Iris dataset with different parameters. Our experiments show that the k-means algorithm is sensitive to initial cluster centers and may terminates at local optimum solutions.

As the most popular and the simplest partitional clustering algorithm, the k-means algorithm has a long history. In fact, the algorithm was independently discovered by several people from different scientific fields [145]. Since then many variations of the k-means algorithm have been proposed. For more information about the k-means algorithm and its variations, readers are referred to [90] and references therein.

9

The c-means Algorithm

The c-means algorithm is also referred to as the fuzzy c-means (FCM) algorithm, which was developed by [67] and improved by [22]. Since the c-means algorithm is a fuzzy clustering algorithm, it allows one record to belong to two or more clusters with some weights. The c-means algorithm is very similar to the k-means algorithm in other aspects. In this chapter, we shall implement the c-means algorithm and illustrate it with some examples.

9.1 Description of the Algorithm

Let $X = \{\mathbf{x}_0, \mathbf{x}_1, \cdots, \mathbf{x}_{n-1}\}$ be a dataset containing n records, each of which is described by d numeric attributes. Let U be a $n \times k$ fuzzy partition matrix, which satisfies the following conditions:

$$0 \le u_{ij} \le 1, \quad 0 \le j \le k-1, 1 \le i \le n-1, \qquad (9.1a)$$

$$\sum_{j=1}^{k} u_{ij} = 1, \quad 0 \le i \le n-1, \qquad (9.1b)$$

$$\sum_{i=1}^{n} u_{ij} > 0, \quad 0 \le j \le k-1, \qquad (9.1c)$$

where u_{ij} is the (i,j) entry of the matrix U.

Given the dataset X, the c-means algorithm finds a fuzzy partition of X by minimizing the following objective function:

$$J_\alpha = \sum_{j=0}^{k-1} \sum_{i=0}^{n-1} u_{ij}^\alpha D_{euc}(\mathbf{x}_i, \boldsymbol{\mu}_j)^2, \qquad (9.2)$$

where $\alpha \in (1, \infty)$ is a weighting exponent, the $\boldsymbol{\mu}_j$ is the center of cluster j, and $D_{euc}(\cdot, \cdot)$ is the Euclidean distance.

To minimize the objective function, the c-means algorithm employs an iterative process. That is, the c-means algorithm repeats updating the fuzzy cluster memberships given the cluster centers and updating the cluster centers given the fuzzy cluster memberships until some stop condition is met. At the beginning, the c-means algorithm selected k distinct records from the dataset as initial cluster centers. Suppose $\boldsymbol{\mu}_0^{(0)}, \boldsymbol{\mu}_1^{(0)}, \cdots, \boldsymbol{\mu}_{k-1}^{(0)}$ are the k initial cluster centers. Then the c-means algorithm updates the fuzzy cluster memberships according to the following formula:

$$u_{ij}^{(0)} = \left[\sum_{l=0}^{k-1} \left(\frac{D_{euc}(\mathbf{x}_i, \boldsymbol{\mu}_j^{(0)})}{D_{euc}(\mathbf{x}_i, \boldsymbol{\mu}_l^{(0)})} \right)^{\frac{2}{\alpha-1}} \right]^{-1}$$
$$= \frac{D_{euc}(\mathbf{x}_i, \boldsymbol{\mu}_j^{(0)})^{\frac{-2}{\alpha-1}}}{\sum_{l=0}^{k-1} D_{euc}(\mathbf{x}_i, \boldsymbol{\mu}_l^{(0)})^{\frac{-2}{\alpha-1}}} \qquad (9.3)$$

for $j = 0, 1, \cdots, k-1$ and $i = 0, 1, \cdots, n-1$.

Once the fuzzy cluster memberships are updated, the c-means continues to update the cluster centers according to the following formula:

$$\boldsymbol{\mu}_j^{(1)} = \frac{\sum_{i=0}^{n-1} \left(u_{ij}^{(0)} \right)^\alpha \mathbf{x}_i}{\sum_{i=0}^{n-1} \left(u_{ij}^{(0)} \right)^\alpha} \qquad (9.4)$$

for $j = 0, 1, \cdots, k-1$.

The c-means algorithm repeats the above steps until the change of the objective function values between two iterations is within the tolerance or the maximum number of iterations is reached.

Once a fuzzy partition U is obtained, the c-means algorithm produces a hard partition based on the fuzzy partition. Precisely, let $\gamma_0, \gamma_1, \cdots, \gamma_{n-1}$ be the hard partition. That is, γ_i is the index of the cluster to which record $\mathbf{x}_i$ belongs. Then the hard partition can be determined as follows:

$$\gamma_i = \operatorname*{argmax}_{0 \le j \le k-1} u_{ij}, \quad i = 0, 1, \cdots, n-1. \qquad (9.5)$$

9.2　Implementation

It is straightforward to implement the c-means algorithm. In Python, we can implement the c-means algorithm as follows:

```
def cmeans(X, k=3, alpha=2, tol=1e-6, maxit=100):
    if alpha <= 1:
        raise ValueError("Invalid alpha")
    X = np.ascontiguousarray(X)
    n, d = X.shape
    epsilon = 1e-8
```

```
7     ind = np.random.choice(n, k, replace=False)
8     clusterCenters = X[ind,:]
9     dm = np.zeros((n,k))
10    for i in range(k):
11        dm[:,i] = np.sum(np.square(X-clusterCenters[i,:]),
              axis=1)
12    dma = np.pow(dm + epsilon, -1/(alpha-1))
13    clusterMembership = dma / np.sum(dma, axis=1, keepdims=
          True)
14    objectiveValue = np.sum(np.multiply(np.pow(
          clusterMembership, alpha), dm))
15    numIter = 1
16    while numIter < maxit:
17        # update cluster centers
18        ua = np.pow(clusterMembership, alpha)
19        for i in range(k):
20            clusterCenters[i,:] = np.sum(X*ua[:,i].reshape
                  ((n,1)), axis=0)/np.sum(ua[:,i])
21        # update cluster membership
22        objectiveValue_ = objectiveValue
23        for i in range(k):
24            dm[:,i] = np.sum(np.square(X-clusterCenters[i
                  ,:]), axis=1)
25        dma = np.pow(dm + epsilon, -1/(alpha-1))
26        clusterMembership = dma / np.sum(dma, axis=1,
              keepdims=True)
27        objectiveValue = np.sum(np.multiply(np.pow(
              clusterMembership, alpha), dm))
28        numIter += 1
29        if np.abs(objectiveValue - objectiveValue_) < tol:
30            break;
31    return clusterMembership, clusterCenters,
          objectiveValue.item(), numIter
```

In the above code, we try to use vectorized operations as much as possible. Initialization of the cluster centers is the same as in the k-means algorithm. We select k distinct records as the initial cluster centers. Cluster memberships are calculated according to Equation (9.3).

During the iterative process, we update the cluster centers and the cluster memberships alternatively. To prevent the division by zero error, we add a small positive number of the distances when updating the cluster memberships.

In our implementation, the distance matrix dm stores the squared Euclidean distances between all records and all cluster centers. When updating the cluster memberships, we use the power $\dfrac{-1}{\alpha - 1}$ rather than the power $\dfrac{-2}{\alpha - 1}$. Such treatment can save some numerical operations.

Since the *c*-means algorithm also uses random initial cluster centers, we can run the *c*-means algorithm multiple times to get the best run. We can implement the multiple run as follows:

```
def cmeans2(X, k=3, alpha=2, numrun=10, maxit=100):
    bestFM, bestCC, bestOV, bestIters = cmeans(X, k=k,
        alpha=alpha, maxit=maxit)
    print([bestOV, bestIters])
    for i in range(numrun-1):
        fm, cc, ov, iters = cmeans(X, k=k, alpha=alpha,
            maxit=maxit)
        print([ov, iters])
        if ov < bestOV:
            bestFM, bestCC, bestOV, bestIters = fm, cc, ov,
                iters
    return bestFM, bestCC, bestOV, bestIters
```

Given the same parameters, we select the run with the lowest objective function value as the best run.

9.3   Examples

In this section, we apply the *c*-means algorithm implemented in the previous section to cluster a synthetic dataset and the Iris dataset.

The following piece of code shows how to apply the *c*-means algorithm to a synthetic dataset multiple times:

```
centers = [[3, 3], [-3, -3], [3, -3]]
X, y = make_blobs(n_samples=300, centers=centers,
    cluster_std=1, random_state=1)

fm, cc, ov, iters = cmeans2(X, k=3)
yhat = np.argmax(fm, axis=1)
cm1 = createCM(y, yhat)
print([ov, iters])
print(cm1)
print(np.array_str(fm[::15,:], precision=4, suppress_small=
    True))
```

After executing the above block of code, we see the following output:

```
[490.91547109802275, 12]
[490.9154711285173, 11]
[490.91547108445354, 12]
[490.915471116234, 8]
[490.91547109508303, 8]
```

```
 6  [490.91547108554346,  11]
 7  [490.9154710835343,  9]
 8  [490.9154710813908,  13]
 9  [490.9154711012434,  8]
10  [490.915471083445,  9]
11  [490.9154710813908,  13]
12          0    1    2
13  0   100    0    0
14  1     0    1   99
15  2     1   99    0
16  [[0.0041 0.0072 0.9886]
17   [0.735  0.211  0.054 ]
18   [0.0155 0.0395 0.945 ]
19   [0.0059 0.9885 0.0056]
20   [0.0222 0.0418 0.936 ]
21   [0.0513 0.9016 0.0471]
22   [0.955  0.0301 0.015 ]
23   [0.9003 0.0746 0.025 ]
24   [0.0197 0.96   0.0203]
25   [0.0333 0.916  0.0507]]
```

From the first part of the output, we see that the c-means algorithm achieved almost the same objective function values in all ten runs. This indicates that the c-means algorithm is less sensitive to initial cluster centers than the k-means algorithm. From the confusion matrix, we see that only two records were misclassified. The last part of the output shows the fuzzy memberships of selected records.

In the above example, we used $\alpha = 2$ in the algorithm. To see the impact of α, let us run the algorithm with $\alpha = 8$ by executing the following block of code:

```
1  fm, cc, ov, iters = cmeans2(X, k=3, alpha=8)
2  yhat = np.argmax(fm, axis=1)
3  cm1 = createCM(y, yhat)
4  print([ov, iters])
5  print(cm1)
6  print(np.array_str(fm[::30,:], precision=4, suppress_small=
       True))
```

The output is

```
1  [1.609246300468918,  15]
2  [1.6092461622020275,  18]
3  [1.6092461891446623,  21]
4  [1.6092462157840335,  15]
5  [1.609246344043309,  17]
6  [1.6092464432186513,  19]
7  [1.6092462255305717,  14]
```

```
 8 [1.609246420416092, 24]
 9 [1.6092461775655393, 18]
10 [1.6092463656460998, 20]
11 [1.6092461622020275, 18]
12      0    1    2
13 0   0   100   0
14 1   1    0   99
15 2  99    1    0
16 [[0.2635 0.2405 0.496 ]
17  [0.3314 0.3987 0.2699]
18  [0.2877 0.2489 0.4634]
19  [0.6351 0.1832 0.1817]
20  [0.2899 0.2612 0.4489]
21  [0.4554 0.2742 0.2704]
22  [0.2819 0.4658 0.2523]
23  [0.3048 0.4352 0.26  ]
24  [0.4418 0.2785 0.2797]
25  [0.4516 0.2659 0.2825]]
```

We see similar results except that the fuzzy memberships are more evenly distributed when α is larger.

To apply the c-means algorithm to the Iris dataset with three clusters, we can run the following code:

```
 1 iris = fetch_ucirepo(id=53)
 2 X = iris.data.features
 3 y = iris.data.targets
 4
 5 fm, cc, ov, iters = cmeans2(X, k=3)
 6 yhat = np.argmax(fm, axis=1)
 7 cm1 = createCM(y, yhat)
 8 print([ov, iters])
 9 print(cm1)
10 print(np.array_str(fm[::30,:], precision=4, suppress_small=
      True))
```

Default values of other parameters are used. Executing the above block of code gives the following output:

```
 1 [60.57595596119823, 20]
 2 [60.575955958487526, 15]
 3 [60.57595583673644, 25]
 4 [60.57595595099188, 28]
 5 [60.57595582822921, 16]
 6 [60.57595583847882, 21]
 7 [60.57595583464465, 18]
 8 [60.57595590575501, 20]
 9 [60.57595592703868, 17]
```

```
10  [60.57595613413543, 43]
11  [60.57595582822921, 16]
12                      0   1   2
13  Iris-setosa         0   0  50
14  Iris-versicolor     3  47   0
15  Iris-virginica     37  13   0
16  [[0.0012 0.0025 0.9963]
17   [0.0062 0.0141 0.9798]
18   [0.1449 0.6356 0.2195]
19   [0.0493 0.931  0.0197]
20   [0.9705 0.0257 0.0038]]
```

We see similar patterns as for the synthetic data. The c-means algorithm is less sensitive to initial cluster centers that the k-means algorithm. From the confusion matrix, we see that 16 records were misclassified.

We can plot the fuzzy memberships by using stacked bar charts. The following block of code illustrates how to create a stacked bar chart from a fuzzy membership matrix:

```
1  x = [10*i for i in range(fm.shape[0]//10)]
2  df = pd.DataFrame(fm[x,:],
3    index=x,
4    columns=["C1", "C2", "C3"])
5  ax = df.plot(
6      kind = 'barh',
7      stacked = True,
8      colormap='Greys',
9      title = 'Fuzzy membership',
10     mark_right = True)
11 fig = ax.get_figure()
12 fig.savefig('irisfm2.pdf')
```

In the above code, we first create a Pandas data frame from a subset of the fuzzy membership matrix `fm` and then use the data frame's plot function to create the stacked bar chart. We only plot a subset of the fuzzy membership matrix to save space.

Figure 9.1 shows the stacked bar charts of fuzzy memberships of selected records. Figure 9.1a shows the fuzzy memberships when $\alpha = 2$. Figure 9.1b shows the fuzzy memberships when $\alpha = 8$. Comparing the two stacked bar charts, we see that the higher the value of α, the more uniform of the fuzzy memberships.

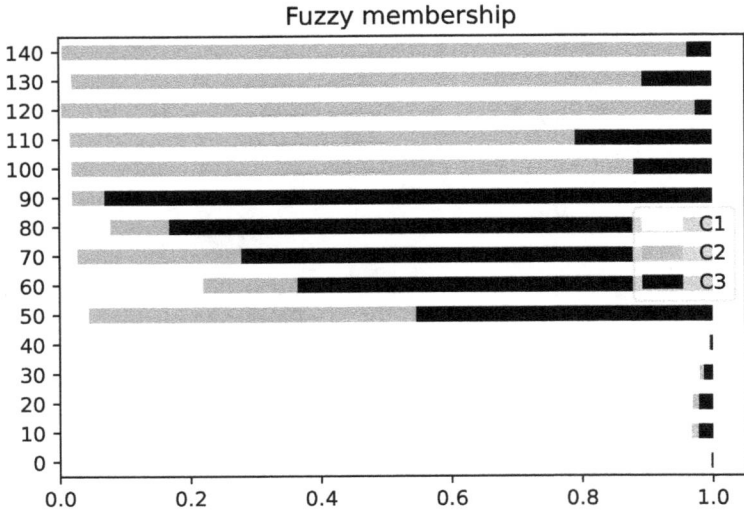

(a) $\alpha = 2$

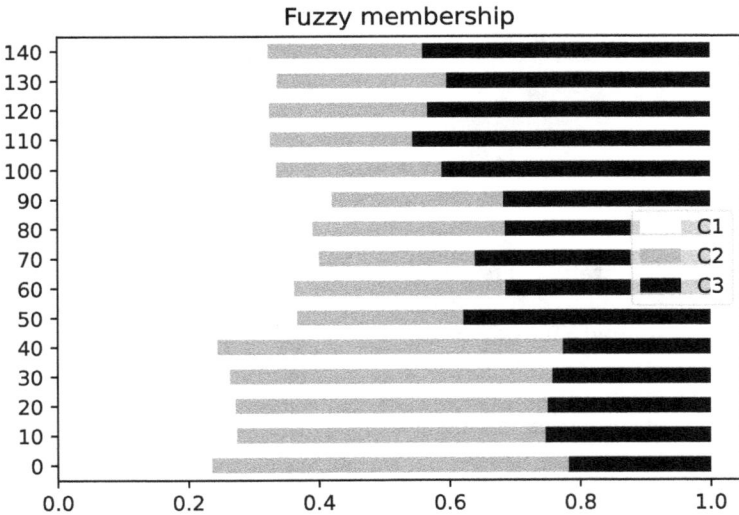

(b) $\alpha = 8$

FIGURE 9.1: Stacked bar charts of fuzzy memberships of selected records.

9.4 Summary

In this chapter, we implemented the c-means algorithm and illustrated the algorithm with several examples. The c-means algorithm implemented in this chapter is very similar to the fuzzy k-means algorithm [21, 100]. The c-means algorithm is one of the many fuzzy cluster algorithms. For more information about fuzzy clustering, readers are referred to [130], [184], and [243].

10

The *k*-prototypes Algorithm

The *k*-prototypes algorithm [136] is a clustering algorithm designed to cluster mixed-type datasets. The *k*-prototypes algorithm was developed based on the idea of the *k*-means algorithm and the *k*-modes algorithm [41, 136]. In this chapter, we shall implement the *k*-prototypes algorithm.

10.1   Description of the Algorithm

Let $X = \{\mathbf{x}_0, \mathbf{x}_1, \cdots, \mathbf{x}_{n-1}\}$ be a mixed-type dataset containing n records, each of which is described by d attributes. Suppose that p attributes are numerical and the rest $d-p$ attributes are categorical. Without loss of generality, we assume that the first p attributes are numeric and the last $d-p$ attributes are categorical. Then the distance between two records $\mathbf{x}$ and $\mathbf{y}$ in X can be defined as

$$D_{mix}(\mathbf{x}, \mathbf{y}, \beta) = \sum_{h=0}^{p-1}(x_h - y_h)^2 + \beta \sum_{h=p}^{d-1}\delta(x_h, y_h), \qquad (10.1)$$

where x_h and y_h are the hth component of $\mathbf{x}$ and $\mathbf{y}$, respectively, β is a balance weight used to avoid favoring either type of attribute, and $\delta(\cdot, \cdot)$ is the simple matching distance defined as

$$\delta(x_h, y_h) = \begin{cases} 0, & \text{if } x_h = y_h, \\ 1, & \text{if } x_h \neq y_h. \end{cases}$$

The objective function that the *k*-prototypes algorithm tries to minimize is defined as

$$P_\beta = \sum_{j=0}^{k-1} \sum_{\mathbf{x} \in C_j} D_{mix}(\mathbf{x}, \boldsymbol{\mu}_j, \beta), \qquad (10.2)$$

where $D_{mix}(\cdot, \cdot, \beta)$ is defined in Equation (10.1), k is the number of clusters, C_j is the jth cluster, and $\boldsymbol{\mu}_j$ is the center or prototype of cluster C_j.

To minimize the objective function defined in Equation (10.2), the algorithm proceeds iteratively. That is, the *k*-prototypes algorithm repeats updating the cluster memberships given the cluster centers and updating the cluster centers given the cluster memberships until some stop condition is met.

DOI: 10.1201/9781003592648-10   148

At the beginning, the k-prototypes algorithm initializes the k cluster centers by selecting k distinct records from the dataset randomly. Suppose $\boldsymbol{\mu}_0^{(0)}, \boldsymbol{\mu}_1^{(0)}, \cdots, \boldsymbol{\mu}_{k-1}^{(0)}$ are the k initial cluster centers. The k-prototypes algorithm updates the cluster memberships $\gamma_0, \gamma_1, \cdots, \gamma_{n-1}$ according to the following formula:

$$\gamma_i^{(0)} = \operatorname*{argmin}_{0 \le j \le k-1} D_{mix}(\mathbf{x}_i, \boldsymbol{\mu}_j^{(0)}, \beta), \tag{10.3}$$

where $D_{mix}(\cdot, \cdot, \beta)$ is defined in Equation (10.1).

Once the cluster memberships are updated, the algorithm continues to update the cluster centers according to the following formula:

$$\mu_{jh}^{(1)} = \frac{1}{|C_j|} \sum_{\mathbf{x} \in C_j} x_h, \quad h = 0, 1, \cdots, p-1, \tag{10.4a}$$

$$\mu_{jh}^{(1)} = \operatorname{mode}_h(C_j), \quad h = p, p+1, \cdots, d-1, \tag{10.4b}$$

where $C_j = \left\{ \mathbf{x}_i \in X : \gamma_i^{(0)} = j \right\}$ for $j = 0, 1, \cdots, k-1$, and $\operatorname{mode}_h(C_j)$ is the most frequent categorical value of the hth attribute in cluster C_j. Let $A_{h0}, A_{h1}, \cdots, A_{h,m_h-1}$ be the distinct values the hth attribute can take, where m_h is the number of distinct values the hth attribute can take. Let $f_{ht}(C_j)$ be the number of records in cluster C_j, whose hth attribute takes value A_{ht} for $t = 0, 1, \cdots, m_h - 1$. That is,

$$f_{ht}(C_j) = |\{\mathbf{x} \in C_j : x_h = A_{ht}\}|, \quad , t = 0, 1, \cdots, m_h - 1.$$

Then

$$\operatorname{mode}_h(C_j) = \max_{0 \le t \le m_h - 1} f_{ht}(C_j), \quad h = p, p+1, \cdots, d-1.$$

The k-prototypes algorithm repeats the above steps until the cluster memberships do not change or the maximum number of iterations is reached.

10.2 Implementation

In this section, we implement the k-prototypes algorithm described in the previous section. Since the k-prototypes algorithm is designed to clusterize mixed-type data, we need to think about a way to input the mixed-type data to the algorithm. In [94], a single dataset is inputted to the C++ algorithm and a schema is used to tell which columns are numerical and which columns are categorical. Since Python is a scripting language, following the C++ approach can affect the performance. In the Python program, we will input the numerical part and the categorical part of the mixed-type dataset to the algorithm separately.

In Python, we implement the k-prototypes algorithm as follows:

```
def kproto(Xn, Xc, k=3, beta=None, tol=1e-8, maxit=100):
    Xn = np.ascontiguousarray(Xn)
    Xc = np.ascontiguousarray(Xc)
    n, d1 = Xn.shape
    nc, d2 = Xc.shape
    if n != nc:
        raise ValueError("dimension mismatch")
    if beta is None:
        beta = estBeta(Xn, Xc)
    ind = np.random.choice(n, k, replace=False)
    clusterCentersn = Xn[ind,:]
    clusterCentersc = Xc[ind,:]
    dm = np.zeros((n,k))
    for i in range(k):
        dm[:,i] = np.sum(np.square(Xn-clusterCentersn[i,:])
            , axis=1) + beta * np.count_nonzero(Xc-
            clusterCentersc[i,:], axis=1)
    clusterMembership = np.argmin(dm, axis=1)
    objectiveValue = np.sum(dm[list(range(n)),
        clusterMembership]).item()
    numIter = 1
    while numIter < maxit:
        # update cluster centers
        for i in range(k):
            bInd = clusterMembership==i
            if np.any(bInd):
                clusterCentersn[i,:] = np.mean(Xn[bInd],
                    axis=0)
                clusterCentersc[i,:] = stats.mode(Xc[bInd])
                    [0]
            else:
                clusterCentersn[i,:] = Xn[np.random.randint
                    (0, n),:]
                clusterCentersc[i,:] = Xc[np.random.randint
                    (0, n),:]
        # update cluster membership
        for i in range(k):
            dm[:,i] = np.sum(np.square(Xn-clusterCentersn[i
                ,:]), axis=1) + beta * np.count_nonzero(Xc-
                clusterCentersc[i,:], axis=1)
        clusterMembership = np.argmin(dm, axis=1)
        objectiveValue_ = objectiveValue
        objectiveValue = np.sum(dm[list(range(n)),
            clusterMembership]).item()
        numIter += 1
```

```
36    if np.abs(objectiveValue - objectiveValue_) < tol:
37        break;
38  return clusterMembership, clusterCentersn,
      clusterCentersc, objectiveValue, numIter
```

In the above code, the arguments Xn and Xc correspond to the numerical part and the categorical part of the mixed-type dataset. In Lines 6–7, we check the number of rows of Xn and Xc to ensure that they have the same number of records.

In Lines 10–12, we initialize the cluster centers by randomly selecting k distinct records from the input data. In Lines 13–15, we calculate the distances between all records and all cluster centers. Here, vectorized operations are used to speed up the calculation.

During the iterative process, we update the cluster centers and the cluster memberships alternatively. When an empty cluster appears, the cluster center is updated to a randomly selected record. When a cluster is not empty, its center is updated according to the formulas given in the previous section. That is, the numerical part of the center is updated by the mean of all records in the cluster and the categorical part is updated by the mode of all records in the cluster. The mode function from the SciPy library is used to get the mode.

When the parameter β is not specified in the input, the estBeta function is called to estimate the parameter. The estBeta function is given below:

```
1 def estBeta(Xn, Xc):
2     numVar = np.mean(np.var(Xn, axis=0))
3     vv = np.zeros(Xc.shape[1])
4     for j in range(Xc.shape[1]):
5         nv, nc = np.unique(Xi[:,j], return_counts=True)
6         vp = nc/np.sum(nc)
7         vv[j] = 1 - np.sum(np.square(vp))
8     return numVar / np.mean(vv)
```

The method used to estimate the parameter β is a method used in the R package clustMixType [238]. In this method, the parameter β is estimated to be the ratio of the average numerical variance to the average categorical variance.

Since the k-prototypes algorithm used random initial cluster centers, we also need to run the algorithm multiple time to minimize the affect of initial cluster centers. The following function implements the multiple run of the algorithm:

```
1 def kproto2(Xn, Xc, k=3, beta=None, numrun=10, maxit=100):
2     bestCM, bestCCn, bestCCc, bestOV, bestIters = kproto(Xn
      , Xc, k=k, beta=beta, maxit=maxit)
3     vOV = np.zeros(numrun)
4     vOV[0] = bestOV
5     for i in range(numrun-1):
```

```
6  cp, ccn, ccc, ov, iters = kproto(Xn, Xc, k=k, beta=
       beta, maxit=maxit)
7  vOV[i+1] =ov
8  if ov < bestOV:
9      bestCM, bestCCn, bestCCc, bestOV, bestIters =
           cp, ccn, ccc, ov, iters
10 return bestCM, bestCCn, bestCCc, bestOV, bestIters, vOV
```

The best run is determined to be the run with the minimum objective function value.

10.3   Examples

In this section, we apply the k-prototypes algorithm to clusterize a mixed-type dataset from the UCI machine learning repository. The mixed-type dataset we select is the heart disease dataset [11]. This dataset is also used in [94] to illustrate the k-prototypes algorithm.

First, let us load the necessary Python libraries by running the following code:

```
1 import numpy as np
2 import matplotlib.pyplot as plt
3 from sklearn.impute import SimpleImputer
4 from scipy import stats
5 from ucimlrepo import fetch_ucirepo
6 from dcutil import createCM
```

To load the heart disease data, we run the following block of code:

```
1 heart_disease = fetch_ucirepo(id=45)
2
3 X = heart_disease.data.features
4 y = heart_disease.data.targets.copy()
5 y[y>0] = 1
```

The heart disease dataset has a id of 45 in the UCI machine learning repository. The targets of the dataset range from 0 to 4, where 0 indicates no presence of heart disease. In our experiment, we treat records with positive targets as one cluster.

Before we apply the k-prototypes algorithm to the heart disease dataset, we need to preprocess the data as the heart disease dataset contains missing values and its numerical variables have quite different ranges.

To show that the dataset contains missing values, we run the following code:

```
1 print(X.isnull().sum())
```

The output of running the above code is

```
1  age        0
2  sex        0
3  cp         0
4  trestbps   0
5  chol       0
6  fbs        0
7  restecg    0
8  thalach    0
9  exang      0
10 oldpeak    0
11 slope      0
12 ca         4
13 thal       2
14 dtype: int64
```

From the output, we see that the variable `ca` contains four missing values and the variable `thal` contains two missing values.

Since there are only a few missing values, we can impute the missing values. Since both variables take integer values, we use the most frequent values of the variables to replace the missing values. To do that, we can use the `SimpleImputer` from the scikit-learn library. The following block of code shows how this is done:

```
1 imp = SimpleImputer(missing_values=np.nan, strategy='
    most_frequent')
2 imp.fit(X)
3 Xi = imp.transform(X)
```

Now we need to split the heart disease dataset into two parts: the numerical part and the categorical part. The information about which variables are numerical and which variables are categorical can be obtained from the UCI machine learning repository. The following code shows how to split the dataset:

```
1 varNames = list(X.columns)
2 numNames = ['age', 'trestbps', 'chol', 'thalach', 'oldpeak'
    , 'ca']
3 numInd = [varNames.index(s) for s in numNames]
4 catInd = [varNames.index(s) for s in set(varNames).
    difference(numNames)]
5 catInd.sort()
6
7 Xn = Xi[:, numInd]
8 Xc = Xi[:, catInd].astype(int)
```

In the above code, we first obtain the column indices of the numerical variables and the categorical variables. Then we select the numerical part and the categorical part by the corresponding indices.

To prevent one numerical variable to dominate the distance, we normalize all numerical variables by the min-max normalization method. The following code shows how to normalize the numerical data:

```
1  vMin = np.min(Xn, axis=0)
2  vMax = np.max(Xn, axis=0)
3  Xn = (Xn-vMin)/(vMax-vMin)
```

To get the ranges of the preprocessed data and the estimated value of the parameter β, we can run the following code:

```
1  print(np.min(Xn, axis=0))
2  print(np.max(Xn, axis=0))
3  print(np.min(Xc, axis=0))
4  print(np.max(Xc, axis=0))
5  print(estBeta(Xn, Xc))
```

Executing the above block of code in Spyder gives the following output:

```
1  [0. 0. 0. 0. 0. 0.]
2  [1. 1. 1. 1. 1. 1.]
3  [0 1 0 0 0 1 3]
4  [1 4 1 2 1 3 7]
5  0.05851507919628685
```

From the output, we see that the numerical variables were scaled to intervals $[0, 1]$. The categorical variables were coded in integers. The parameter β was estimated to be 0.0585, which can be a good start value for the parameter.

Now we are ready to apply the k-prototypes algorithm to the heart disease dataset. To do that, we run the following code:

```
1  bcm, bccn, bccc, bov, biters, vOV = kproto2(Xn, Xc, k=2,
       numrun=100)
2  cm1 = createCM(y, bcm)
3  print(cm1)
4  print([bov, biters])
```

We run the k-prototypes algorithm 100 times on the heart disease dataset. After executing the above block of code, we see the following output:

```
1        0    1
2  0   14   150
3  1   96    43
4  [91.12791972438707, 8]
```

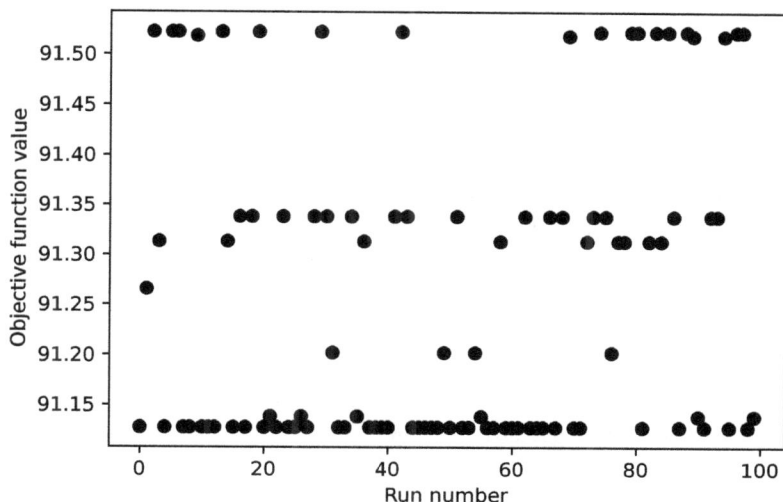

FIGURE 10.1: Objective function values of 100 runs of the k-prototypes algorithm on the heart disease dataset.

From the output, we see that 57 records were misclassified in the best run. The best run produced an objective function value of 91.13. Figure 10.1 shows the objective function values of all the 100 runs. Due to the use of the mismatch distance in the objective function, the objective function values of the 100 runs form clear groups.

To give more weight to the categorical variables in the objective function, we can set the parameter β to be 1, which is about 20 times the estimated value. To run the k-prototypes with $\beta = 1$, we use the following code:

```
bcm, bccn, bccc, bov, biters, vOV = kproto2(Xn, Xc, k=2,
    beta=1, numrun=100)
cm1 = createCM(y, bcm)
print(cm1)
print([bov, biters])
```

Executing the above block of code, we see the following output:

```
0    27    137
1    107    32
[651.6327692076546, 6]
```

The accuracy of the clustering result is about the same as the case when the estimated value of β was used.

If we want to ignore the categorical variables in the objective function, we can set the parameter β to be zero. To test this, we run the following block of code:

```
bcm, bccn, bccc, bov, biters, vOV = kproto2(Xn, Xc, k=2,
    beta=0, numrun=100)
cm1 = createCM(y, bcm)
print(cm1)
print([bov, biters])
```

After executing the above block of code, we see the following output:

```
0   145   19
1    61   78
[46.80906544956523, 7]
```

From the output, we see that the accuracy of the clustering result decreased as more records were misclassified.

10.4 Summary

In this chapter, we implemented the *k*-prototypes algorithm and illustrated it with a real dataset. The *k*-prototypes algorithm integrates the *k*-means algorithm and the *k*-modes algorithm. The distance measure for the *k*-prototypes algorithm includes two components: the distance for numerical attributes and the distance for categorical attributes. The two components are balanced using a weight. In our implementation, we used the squared Euclidean distance for numerical attributes and the simple matching distance for the categorical attributes. In fact, the *k*-prototypes algorithm can use other mixed-type distance measures. For example, the general Minkowski distance [138, 139].

11

The Genetic k-modes Algorithm

The k-modes algorithm is a center-based clustering algorithm designed to cluster categorical datasets [136, 41]. In the k-modes algorithm, centers of clusters are referred to as modes. One drawback of the k-modes algorithm is that the algorithm can only guarantee a locally optimal solution. The genetic k-modes algorithm[93] was developed to improve the k-modes algorithm by integrating the k-modes algorithm with the genetic algorithm [128]. Although the genetic k-modes algorithm can not guarantee a globally optimal solution, the genetic k-modes algorithm has more chances to find a globally optimal solution than the k-modes algorithm. In this chapter, we shall implement the genetic k-modes algorithm and illustrate the algorithm with examples.

11.1 Description of the Algorithm

Before we introduce the genetic k-modes algorithm, let us first give a brief introduction to the genetic algorithm, which was originally introduced by [128]. In genetic algorithms, solutions (i.e., parameters) of a problem are encoded in chromosomes and a population is composed of many solutions. Each solution is associated with a fitness value.

A genetic algorithm evolves over generations. During each generation, three genetic operators, natural selection, crossover, and mutation, are applied to the current population to produce a new population. The natural selection operator selects a few chromosomes from the current population based on the principle of survival of the fittest. Then the selected chromosomes are modified by the crossover operator and the mutation operator before putting into the new population.

The genetic k-modes algorithm was developed based on the genetic k-means algorithm [159]. The genetic k-modes algorithm has five elements: coding, initialization, selection, mutation, and k-modes operator. In the genetic k-modes algorithm, the crossover operator is replaced by the k-modes operator.

DOI: 10.1201/9781003592648-11   157

In the k-modes algorithm, a solution (i.e., a partition) is coded as a vector of cluster memberships. Let $X = \{\mathbf{x}_0, \mathbf{x}_1, \cdots, \mathbf{x}_{n-1}\}$ be a dataset consisting of n records and $C_0, C_1, \cdots, C_{k-1}$ be a partition of X. Then the vector of cluster memberships corresponding to the partition is defined as

$$\gamma_i = j \quad \text{if} \quad \mathbf{x}_i \in C_j$$

for $i = 0, 1, \cdots, n - 1$ and $j = 0, 1, \cdots, k - 1$. A chromosome is said to be legal if each one of the corresponding k clusters is nonempty.

The loss function for a solution represented by $\Gamma = (\gamma_0, \gamma_1, \cdots, \gamma_{n-1})$ is defined as

$$L(\Gamma) = \sum_{i=0}^{n-1} D_{sim}(\mathbf{x}_i, \boldsymbol{\mu}_{\gamma_i}), \tag{11.1}$$

where $\boldsymbol{\mu}_j$ $(0 \le j \le k-1)$ is the center of cluster C_j and $D_{sim}(\cdot, \cdot)$ is the simple matching distance (see Section 5.5.2). The fitness value for the solution is defined as

$$F(\Gamma) = \begin{cases} cL_{max} - L(\Gamma) & \text{if } \Gamma \text{ is legal,} \\ e(\Gamma)F_{min} & \text{otherwise,} \end{cases} \tag{11.2}$$

where $L(\Gamma)$ is defined in Equation (11.1), $c \in (0, 3)$ is a constant, L_{max} is the maximum loss of chromosomes in the current population, F_{min} is the smallest fitness value of the legal chromosomes in the current population if legal chromosomes exist or 1 if otherwise, and $e(\Gamma)$ is the legality ratio defined as the ratio of the number of nonempty clusters over k.

The selection operator randomly selects a chromosome from the current population according to the following distribution:

$$P(\Gamma_i) = \frac{F(\Gamma_i)}{\sum\limits_{r=0}^{N-1} F(\Gamma_i)},$$

where N is the number of chromosomes in the current population and $F(\cdot)$ is defined in Equation (11.2). Hence chromosomes with higher fitness values are more likely to be selected. The selection operator is applied N times in order to select N chromosomes for the new population.

The mutation operator replaces the ith component γ_i of a chromosome $\Gamma = (\gamma_0, \gamma_1, \cdots, \gamma_{n-1})$ with a cluster index randomly selected from $\{0, 1, \cdots, k-1\}$ according to the following distribution:

$$P(j) = \frac{c_m d_{max}(\mathbf{x}_i) - D_{sim}(\mathbf{x}_i, \boldsymbol{\mu}_j)}{\sum\limits_{l=0}^{k-1} [c_m d_{max}(\mathbf{x}_i) - D_{sim}(\mathbf{x}_i, \boldsymbol{\mu}_l)]}, \tag{11.3}$$

where $c_m > 1$ is a constant and

$$d_{max}(\mathbf{x}_i) = \max_{0 \le l \le k-1} D_{sim}(\mathbf{x}_i, \boldsymbol{\mu}_l), \quad i = 0, 1, \cdots, n - 1.$$

In genetic algorithm, mutation occurs with some mutation probability P_m. That is, for each component of each selected chromosome, the mutation operator is applied if a randomly generated standard uniform number is less than P_m.

Once a selected chromosome is mutated, the k-modes operator is applied to it. Let Γ be a chromosome. The k-modes operator changes Γ into another chromosome $\hat{\Gamma}$ based on the following two steps:

Updating cluster centers Given the cluster memberships Γ, the cluster center $\boldsymbol{\mu}_j$ of cluster C_j is updated to the modes of the cluster (see Section 10.1). Let $\hat{\boldsymbol{\mu}}_j$ ($j = 0, 1, \cdots, k-1$) be the updated cluster centers.

Updating cluster memberships Given the cluster centers $\hat{\boldsymbol{\mu}}_j$, the cluster memberships are updated as follows. The ith component γ_i of Γ is updated to

$$\hat{\gamma}_i = \operatorname*{argmin}_{0 \le j \le k-1} D_{sim}(\mathbf{x}_i, \hat{\boldsymbol{\mu}}_j), \quad i = 0, 1, \cdots, n-1.$$

If a cluster is empty, the distance of a record between the cluster is defined to be ∞. Hence illegal chromosomes remain illegal after the application of the k-modes operator.

11.2 Implementation

In this section, we implement the genetic k-modes algorithm in Python. Unlike the k-means and the k-prototypes algorithms, the genetic k-modes algorithm involves operations that cannot be vectorized. The Python code of the genetic k-modes algorithm can be very slow.

The main function of the genetic k-modes algorithm is implemented as follows:

```
def gkmodes(X, k=3, pop=50, maxgen=100, c=1.5, cm=1.5,
    mprob=0.2):
    X = np.ascontiguousarray(X)
    n, d = X.shape
    population = np.random.choice(k, pop*n).reshape(pop, n)
    vLoss = np.zeros((pop, 2))
    vFit = np.zeros(pop)
    for g in range(maxgen):
        # calculate loss, legality ratio
        for i in range(pop):
            vLoss[i,:] = loss(X, population[i,:], k)
        # calculate fitness
```

```
12    bLegal = vLoss[:,1] >= 1
13    maxL = np.max(vLoss[bLegal,0])
14    vFit[bLegal] = c*maxL - vLoss[bLegal,0]
15    minF = np.min(vFit[bLegal])
16    bNotlegal = np.logical_not(bLegal)
17    vFit[bNotlegal] = vLoss[bNotlegal, 1] * minF
18    # selection
19    vP = vFit / np.sum(vFit)
20    ind = np.random.choice(pop, size=pop, p=vP)
21    population = population[ind,:]
22    # mutation
23    for i in range(pop):
24        population[i,:] = mutation(X, population[i,:],
                  k, cm, mprob)
25    # k-modes operator
26    for i in range(pop):
27        population[i,:] = kmode(X, population[i,:], k)
28  for i in range(pop):
29      vLoss[i,:] = loss(X, population[i,:], k)
30  return population, vLoss
```

In the above function, the population is structured as a two-dimensional NumPy array. During the iterations, the fitness values of all chromosomes are first calculated. The selection, the mutation, and the k-modes operators are followed. After a specified number of iterations, the final population and the corresponding loss function values are returned.

The main function depends on three other functions: loss, mutation, and kmode. The loss function is used to calculate the loss function value and the legality ratio of a chromosome. This function is implemented as follows:

```
1  def loss(X, gamma, k):
2      dLoss = 0
3      nNonempty = 0
4      for i in range(k):
5          bInd = gamma == i
6          if np.any(bInd):
7              center = stats.mode(X[bInd])[0]
8              dLoss += np.count_nonzero(X-center)
9              nNonempty += 1
10     return (dLoss, nNonempty/k)
```

The simple matching distance is calculated by counting the non zeros of the difference between a record and its center.

The mutation function is used to perform the mutation operation on a chromosome. This function is implemented as follows:

```
1  def mutation(X, gamma, k, cm, mprob):
2      n = X.shape[0]
3      dm = np.zeros((n, k))
```

```
4      for i in range(k):
5          bInd = gamma == i
6          if np.any(bInd):
7              center = stats.mode(X[bInd])[0]
8          else:
9              center = X[np.random.randint(0, n),:]
10         dm[:,i] = np.count_nonzero(X-center, axis=1)
11     dm = cm*np.max(dm, axis=1, keepdims=True) - dm
12     dm = dm / np.sum(dm, axis=1, keepdims=True)
13     ind = np.where( np.random.rand(n) < mprob )[0]
14     clusterMembership = gamma
15     for i in ind:
16         clusterMembership[i] = np.random.choice(k, size=1,
                   p=dm[i,:].flatten()).item()
17     return clusterMembership
```

The mutation operator cannot be vectorized as the distributions used to select cluster indices are different for different records.

The k-modes operator is implemented as follows:

```
1  def kmode(X, gamma, k):
2      n = X.shape[0]
3      dm = np.zeros((n, k))
4      for i in range(k):
5          bInd = gamma == i
6          if np.any(bInd):
7              center = stats.mode(X[bInd])[0]
8          else:
9              center = X[np.random.randint(0, n),:]
10         dm[:,i] = np.count_nonzero(X-center, axis=1)
11     return np.argmin(dm, axis=1)
```

In the above function, the input is a vector of cluster memberships. The k-modes operator first determines the center (i.e., mode) of each cluster and then assign each record to its nearest center.

11.3   Examples

In this section, we apply the genetic k-modes algorithm to a real dataset. Since the algorithm is developed for categorical data, we use the small soybean dataset [210] from the UCI machine learning repository.

Before we run the code given in the previous section, we need to load the necessary libraries by running the following code:

```
1 import time
2 import numpy as np
3 from scipy import stats
4 from ucimlrepo import fetch_ucirepo
5 from dcutil import createCM
```

To load the small soybean dataset, we run the following code:

```
1 soybean_small = fetch_ucirepo(id=91)
2
3 X = soybean_small.data.features
4 y = soybean_small.data.targets
```

To apply the genetic k-modes algorithm to the soybean dataset with four clusters, we use the following code:

```
1  begt = time.time()
2  cms, losses = gkmodes(X, k=4)
3  endt = time.time()
4  print(endt-begt)
5
6  bestInd = np.argmin(losses[:,0])
7  print(bestInd)
8  yhat = cms[bestInd,:]
9  cm1 = createCM(y, yhat)
10 print(cm1)
```

Default values of other parameters are used. We also measure the runtime of the function. Executing the above block of code, we see the following output:

```
1  332.85249972343445
2  1
3         0    1    2    3
4  D1     0    0   10    0
5  D2     0    0    0   10
6  D3    10    0    0    0
7  D4     1   16    0    0
```

From the output, we see that it took the Python program about 332.85 seconds to finish the algorithm. It only took the C++ program [94] about 10 seconds to do the same calculation. We can see that Python can be very slow for loops. From the confusion matrix, we see that only one record was misclassified by the genetic k-modes algorithm.

To apply the algorithm with three clusters, we use the following code:

```
1 begt = time.time()
2 cms2, losses2 = gkmodes(X, k=3)
3 endt = time.time()
4 print(endt-begt)
5
```

```
 6  bestInd2 = np.argmin(losses2[:,0])
 7  print(bestInd2)
 8  yhat2 = cms2[bestInd2,:]
 9  cm1 = createCM(y, yhat2)
10  print(cm1)
```

Executing the above block of code, we see the following output:

```
1  304.3224980831146
2  0
3        0   1    2
4  D1    0   10   0
5  D2    0   0    10
6  D3    10  0    0
7  D4    17  0    0
```

From the output, we see that it took the Python program about 304.32 seconds to finish the computation.

11.4   Summary

In this chapter, we implemented the genetic k-modes algorithm. We also illustrated the algorithm with examples and tested the sensitivity of the mutation probability. The genetic k-modes algorithm increases the clustering accuracy of the k-modes algorithm. However, the running time of the genetic k-modes algorithm is significantly longer than that of the k-modes algorithm. The Python implementation is about 30 times slower than the C++ implementation given in [94].

The genetic k-modes algorithm implemented in this chapter is one of the search-based clustering algorithms. Other search-based clustering algorithms include the genetic k-means algorithm (GKA) [159] and clustering algorithms based on the tabu search method [90].

12

The FSC Algorithm

In fuzzy clustering algorithms such as the c-means algorithm, each record has a fuzzy membership associated with each cluster that indicates the degree of association of the record to the cluster. In the fuzzy subspace clustering (FSC) algorithm, each attribute has a fuzzy membership associated with each cluster that indicates the degree of importance of the attribute to the cluster. In this chapter, we shall implement the FSC algorithm [89, 90].

12.1   Description of the Algorithm

The FSC algorithm is an extension of the k-means algorithm for subspace clustering. The FSC algorithm imposes weights on the distance measure of the k-means algorithm. Given a dataset $X = \{\mathbf{x}_0, \mathbf{x}_1, \cdots, \mathbf{x}_{n-1}\}$ consisting of n records, each of which is described by d numeric attribute. Recall that the objective function of the k-means algorithms is

$$E = \sum_{j=0}^{k-1} \sum_{\mathbf{x} \in C_j} D_{euc}(\mathbf{x}, \boldsymbol{\mu}_j)^2 = \sum_{j=0}^{k-1} \sum_{\mathbf{x} \in C_j} \sum_{r=0}^{d-1} (x_r - \mu_{jr})^2,$$

where $C_0, C_1, \cdots, C_{k-1}$ are k clusters, $\boldsymbol{\mu}_j$ $(0 \leq j \leq k-1)$ is the center of cluster C_j, $D_{euc}(\cdot, \cdot)$ is the Euclidean distance, and x_r and μ_{jr} are the rth components of $\mathbf{x}$ and $\boldsymbol{\mu}_j$, respectively.

The objective function of the FSC algorithm is defined as

$$E_{\alpha,\epsilon} = \sum_{j=0}^{k-1} \sum_{\mathbf{x} \in C_j} \sum_{r=0}^{d-1} w_{jr}^\alpha (x_r - \mu_{jr})^2 + \epsilon \sum_{j=0}^{k-1} \sum_{r=0}^{d-1} w_{jr}^\alpha, \qquad (12.1)$$

where $\alpha \in (1, \infty)$ is a weight component or fuzzifier, ϵ is a very small positive real number used to prevent divide-by-zero error, and w_{jr} $(0 \leq j \leq k-1, 0 \leq r \leq d-1)$ is the (j, r) entry of the so called fuzzy dimension weight matrix W. A $k \times d$ weight matrix W satisfies the following conditions:

$$w_{jr} \in [0, 1], \quad 0 \leq j \leq k-1, 0 \leq r \leq d-1, \qquad (12.2\text{a})$$

DOI: 10.1201/9781003592648-12

$$\sum_{r=0}^{d-1} w_{jr} = 1, \quad 0 \le j \le k-1. \tag{12.2b}$$

The FSC algorithm tries to minimize the objective function defined in Equation (12.1) using an iterative process. The iterative process is very similar to that of the k-means algorithm. That is, the FSC algorithm repeats updating the cluster centers given the fuzzy dimension weight matrix and updating the fuzzy dimension weight matrix given the cluster centers.

At the beginning, the FSC algorithm initializes the cluster centers by selecting k distinct records randomly and initializes the fuzzy dimension weight matrix equally, i.e., $w_{jr}^{(0)} = \frac{1}{d}$ for $j = 0, 1, \cdots, k-1$, $r = 0, 1, \cdots, d-1$. Suppose $\boldsymbol{\mu}_j^{(0)}$ ($j = 0, 1, \cdots, k-1$) are the initial cluster centers. Then the FSC algorithm updates the clusters $C_0, C_1, \cdots, C_{k-1}$ based on the initial cluster centers and the initial fuzzy dimension weight matrix as follows:

$$C_j^{(0)} = \left\{ \mathbf{x} \in X : D_{euc}\left(\mathbf{x}, \boldsymbol{\mu}_j^{(0)}, W^{(0)}\right) = \min_{0 \le l \le k-1} D_{euc}\left(\mathbf{x}, \boldsymbol{\mu}_l^{(0)}, W^{(0)}\right) \right\}, \tag{12.3}$$

for $j = 0, 1, \cdots, k-1$, where

$$D_{euc}\left(\mathbf{x}, \boldsymbol{\mu}_l^{(0)}, W^{(0)}\right) = \sum_{r=0}^{d-1} \left(w_{lr}^{(0)}\right)^\alpha \left(x_r - \mu_{lr}^{(0)}\right)^2, \quad 0 \le l \le k-1. \tag{12.4}$$

Then the FSC algorithm updates the fuzzy dimension weight matrix based on the cluster centers and the clusters according to the following formula:

$$w_{jr}^{(1)} = \left[\sum_{l=0}^{d-1} \left(\frac{\sum\limits_{\mathbf{x} \in C_j^{(0)}} \left(x_r - \mu_{jr}^{(0)}\right)^2 + \epsilon}{\sum\limits_{\mathbf{x} \in C_j^{(0)}} \left(x_l - \mu_{jl}^{(0)}\right)^2 + \epsilon} \right)^{\frac{1}{\alpha-1}} \right]^{-1}$$

$$= \frac{\left(\sum\limits_{\mathbf{x} \in C_j^{(0)}} \left(x_r - \mu_{jr}^{(0)}\right)^2 + \epsilon \right)^{-\frac{1}{\alpha-1}}}{\sum\limits_{l=0}^{d-1} \left(\sum\limits_{\mathbf{x} \in C_j^{(0)}} \left(x_l - \mu_{jl}^{(0)}\right)^2 + \epsilon \right)^{-\frac{1}{\alpha-1}}} \tag{12.5}$$

for $j = 0, 1, \cdots, k-1$ and $r = 0, 1, \cdots, d-1$.

Once the fuzzy dimension weight matrix is updated. The FSC algorithm will update the cluster centers based on the fuzzy dimension weight matrix and the clusters according to the following formula:

$$\mu_{jr}^{(1)} = \frac{\sum\limits_{\mathbf{x} \in C_j^{(0)}} x_r}{|C_j^{(0)}|} \tag{12.6}$$

for $j = 0, 1, \cdots, k-1$ and $r = 0, 1, \cdots, d-1$.

The FSC algorithm repeats the above three steps until the change of the objective function between two iterations is small or the maximum number of iterations is reached.

12.2 Implementation

The implementation of the FSC algorithm is similar to those of the k-means algorithm and the c-means algorithm. The following code implements the FSC algorithm:

```
def fsc(X, k=3, alpha=2, tol=1e-6, maxit=100):
    if alpha <= 1:
        raise ValueError("Invalid alpha")
    X = np.ascontiguousarray(X)
    n, d = X.shape
    epsilon = 1e-8
    ind = np.random.choice(n, k, replace=False)
    clusterCenters = X[ind,:]
    featureWeight = np.ones((k, d)) / d
    dm = np.zeros((n, k))
    for i in range(k):
        dm[:,i] = np.sum(np.multiply(np.square(X-
            clusterCenters[i,:]), d**(-alpha)), axis=1)
    clusterMembership = np.argmin(dm, axis=1)
    objectiveValue = np.sum(dm[list(range(n)),
        clusterMembership]).item()
    numIter = 1
    while numIter < maxit:
        # update feature weight
        for i in range(k):
            bInd = clusterMembership==i
            if np.any(bInd):
                dv = np.pow(np.sum(np.square(X[bInd,:]-
                    clusterCenters[i,:])+epsilon, axis=0),
                    -1/(alpha-1))
                featureWeight[i,:] = dv / np.sum(dv)
            else:
                featureWeight[i,:] = 1/d
        # update cluster centers
        for i in range(k):
            bInd = clusterMembership==i
            if np.any(bInd):
                clusterCenters[i,:] = np.mean(X[bInd], axis
                    =0)
```

```
30              else:
31                  clusterCenters[i,:] = X[np.random.randint
                        (0, n),:]
32          # update cluster membership
33          for i in range(k):
34              dm[:,i] = np.sum(np.multiply(np.square(X-
                    clusterCenters[i,:]), featureWeight[
                    clusterMembership,:]**alpha), axis=1)
35          clusterMembership = np.argmin(dm, axis=1)
36          objectiveValue_ = objectiveValue
37          objectiveValue = np.sum(dm[list(range(n)),
                clusterMembership]).item()
38          numIter += 1
39          if np.abs(objectiveValue-objectiveValue_) < tol:
40              break;
41      return clusterMembership, clusterCenters, featureWeight
            , objectiveValue, numIter
```

The cluster centers are initialized similarly as in the k-means algorithm and the c-means algorithm. The feature weights are initialized to be $1/d$, where d is the number of features.

At each iterative process, we update the feature weights, the cluster centers, and the cluster memberships. We calculate the objective function value and compare it with the previous value. If the absolute difference between two consecutive values is less than the tolerance, we terminate the iterative process. Vectorized operations are used to improve performance.

Since the FSC algorithm also uses random initial cluster centers, we can run the algorithm multiple times to get the best run. The following code implements the multiple run of the algorithm:

```
1 def fsc2(X, k=3, alpha=2, numrun=10, maxit=100):
2     bestCM, bestCC, bestFW, bestOV, bestIters = fsc(X, k=k,
          alpha=alpha, maxit=maxit)
3     print([bestOV, bestIters])
4     for i in range(numrun-1):
5         cm, cc, fw, ov, iters = fsc(X, k=k, alpha=alpha,
              maxit=maxit)
6         print([ov, iters])
7         if ov < bestOV:
8             bestCM, bestCC, bestFW, bestOV, bestIters = cm,
                  cc, fw, ov, iters
9     return bestCM, bestCC, bestFW, bestOV, bestIters
```

12.3   Examples

In this section, we apply the FSC algorithm implemented in the previous section to a synthetic dataset and the Iris dataset.

Before proceeding to apply the FSC algorithm, we first load the necessary libraries by executing the following block of code:

```
import numpy as np
import matplotlib.pyplot as plt
from ucimlrepo import fetch_ucirepo
from dcutil import createCM
from kmeans import kmeans2
```

The Python module `dcutil` contains the function `createCM` defined in Listing 6.1. We can just import the function from the module without copying the code of the function. We also import the function `kmeans2` from the module `kmeans`.

To illustrate the FSC algorithm's performance, we create a synthetic dataset with clusters embedded in subspaces. We can create such a dataset as follows:

```
np.random.seed(1)
X = np.zeros((300, 3))
y = np.zeros(300, dtype=int)
for i in range(3):
    ind = [100*i+j for j in range(100)]
    X[ind,:] = np.random.rand(100,3)*2 + 4*i
    X[ind,i] = np.random.rand(100)*12
    y[ind] = i
ind = np.random.permutation(list(range(300)))
X = X[ind,:]
y = y[ind]
```

Executing the above block of code will create a dataset with three subspace clusters. We can plot the dataset as follows:

```
fig = plt.figure(figsize=(5, 5))
ax = fig.add_subplot(projection='3d')
ax.scatter(X[:,0], X[:,1],X[:,2], color="k", s=12)
ax.set_xlabel('x')
ax.set_ylabel('y')
ax.set_zlabel('z')
plt.savefig("3d.pdf", bbox_inches='tight')
```

Figure 12.1 shows the resulting dataset.

To apply the FSC algorithm to the synthetic dataset, we run the following block of code:

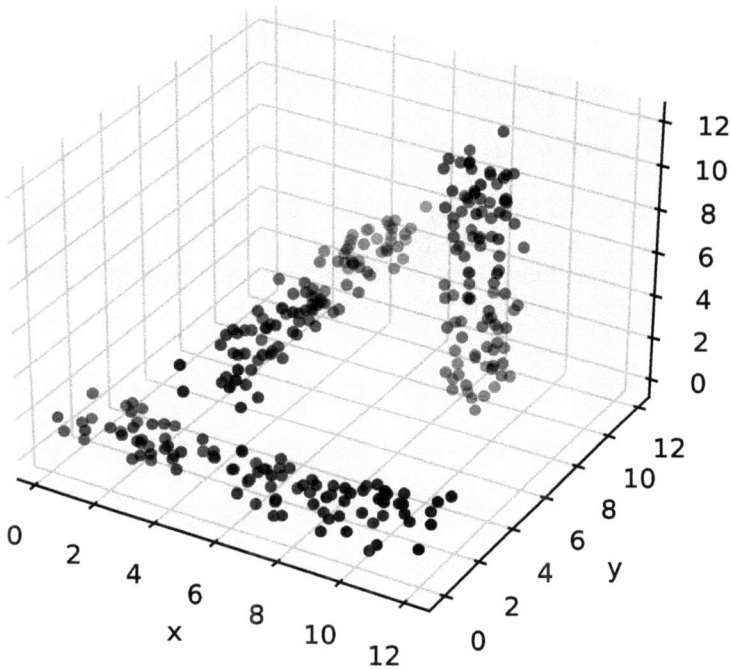

FIGURE 12.1: A synthetic dataset with three subspace clusters.

```
yhat, cc, fw, ov, iters = fsc2(X, k=3, alpha=2)
cm1 = createCM(y, yhat)
print([ov, iters])
print(cm1)
print(np.array_str(fw, precision=4, suppress_small=True))
```

After executing the above code, we see the following output:

```
[144.9954802577933, 6]
[134.7169446057925, 17]
[227.57681283348646, 6]
[109.39339832420217, 9]
[200.07942734318306, 11]
[229.05591137401333, 100]
[136.70664242925852, 9]
[243.740544549417, 7]
[126.38228696270194, 10]
[109.39339832420217, 8]
```

```
11  [109.39339832420217, 9]
12          0      1      2
13  0       0    100      0
14  1       0      0    100
15  2      50     25     25
16  [[0.4332  0.4838  0.083 ]
17   [0.041   0.0421  0.9169]
18   [0.1319  0.0402  0.8279]]
```

From the output, we see that two subspace clusters are correctly identified.
One subspace cluster is not fully recovered. The subspace dimensions of the
clusters are correctly identified by the feature weights. Figure 12.2 shows the
clustering results obtained by the FSC algorithm.

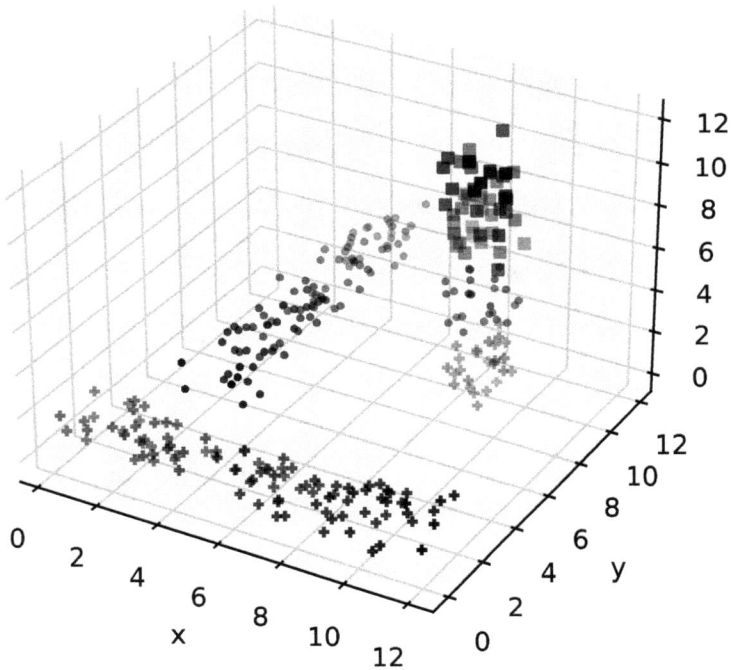

FIGURE 12.2: Clusters obtained by the FSC algorithm.

We can apply the k-means algorithm multiple times to the synthetic
dataset as follows:

```
1  yhat, bcc, bov, biters = kmeans2(X)
2  cm2 = createCM(y, yhat)
3  print([bov, biters])
4  print(cm2)
```

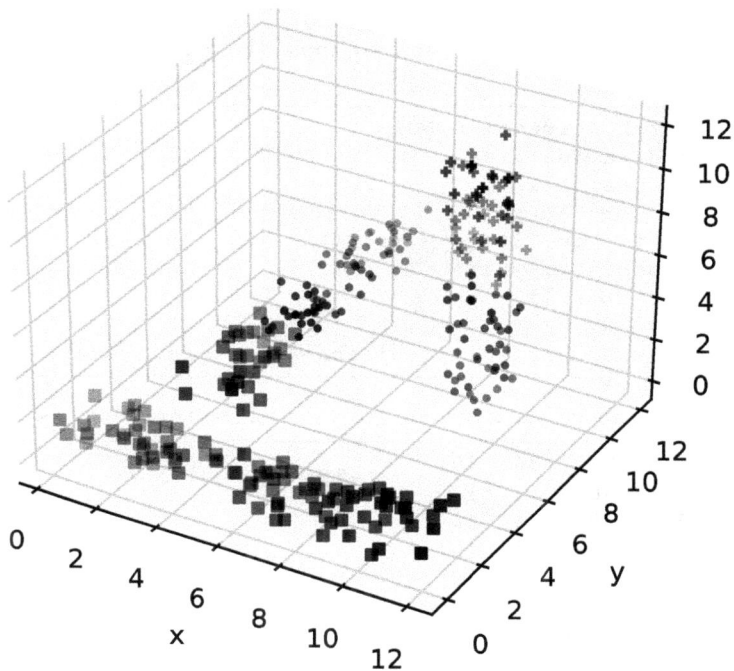

FIGURE 12.3: Clusters obtained by the k-means algorithm.

We see the following output:

```
[3272.75753139638 , 6]
[3307.157048793695 , 11]
[3093.6628681540224 , 5]
[3045.743807105996 , 5]
[3362.2320027642445 , 5]
[3093.6628681540224 , 6]
[3093.6628681540224 , 8]
[3392.207762740887 , 9]
[3045.743807105996 , 5]
[3045.743807105996 , 5]
[3045.743807105996 , 5]
          0    1     2
0    0     0    100
1    0    69    31
2   53    47     0
```

The resulting confusion matrix shows that two subspace clusters are not fully recovered. Figure 12.3 shows the clustering results obtained by the k-means algorithm.

The following code shows how to apply the FSC algorithm to the Iris dataset:

```
1  iris = fetch_ucirepo(id=53)
2  X = iris.data.features
3  y = iris.data.targets
4
5  yhat, cc, fw, ov, iters = fsc2(X, k=3, alpha=3)
6  cm1 = createCM(y, yhat)
7  print([ov, iters])
8  print(cm1)
9  print(np.array_str(fw, precision=4, suppress_small=True))
```

Executing the above block of code gives the following output:

```
1  [0.8286676232227034, 10]
2  [0.888924195786221, 8]
3  [0.8286676232227034, 16]
4  [0.8286676232227034, 10]
5  [0.888924195786221, 8]
6  [0.888924195786221, 9]
7  [1.5975066713188142, 5]
8  [0.8286676232227034, 7]
9  [0.8286676232227034, 14]
10 [0.8286676232227034, 14]
11 [0.8286676232227034, 10]
12                     0    1    2
13 Iris-setosa         0   50    0
14 Iris-versicolor    48    0    2
15 Iris-virginica      4    0   46
16 [[0.1544 0.2564 0.1583 0.4309]
17  [0.138  0.1277 0.2804 0.4538]
18  [0.1523 0.3082 0.1652 0.3743]]
```

The resulting confusion matrix shows that only six records were misclassified by the FSC algorithm.

12.4 Summary

In this chapter, we implemented the FSC algorithm and illustrated the algorithm with several examples. The FSC algorithm is an extension of the k-means algorithm for subspace clustering. The FSC algorithm also applies the idea of fuzzy sets to attribute selection. Rather than treating an attribute is relevant or not relevant to a cluster, the FSC algorithm assigns a weight to the attribute to indicate the importance of the attribute. More information about the FSC algorithm and other relevant algorithms (e.g., mean shift for subspace clustering) can be found in [89] and [90].

13

The Gaussian Mixture Algorithm

Clustering based on Gaussian mixture models is a classical and powerful approach. [38] summarized sixteen Gaussian mixture models, which result in sixteen clustering algorithms. These sixteen Gaussian mixture models are based on different assumptions on the component variance matrices. Four commonly used Gaussian mixture models are [38]:

(a) No restriction is imposed on the component variance matrices $\Sigma_0, \Sigma_1, \cdots,$ and Σ_{k-1};

(b) $\Sigma_0 = \Sigma_1 = \cdots = \Sigma_{k-1} = \Sigma$;

(c) $\Sigma_0 = \Sigma_1 = \cdots = \Sigma_{k-1} = \text{Diag}(\sigma_0^2, \sigma_1^2, \cdots, \sigma_{d-1}^2)$, where $\sigma_0, \sigma_1, \cdots, \sigma_{d-1}$ are unknown;

(d) $\Sigma_0 = \Sigma_1 = \cdots = \Sigma_{k-1} = \text{Diag}(\sigma^2, \sigma^2, \cdots, \sigma^2)$, where σ is unknown.

In this chapter, we implement the clustering algorithm based on the first Gaussian mixture model, i.e., the most general one.

13.1 Description of the Algorithm

Let $X = \{\mathbf{x}_0, \mathbf{x}_1, \cdots, \mathbf{x}_{n-1}\}$ be a numeric dataset containing n records, each of which is described by d numeric attributes. In Gaussian mixture models, each record in the dataset X is assumed to be a sample drawn from a distribution characterized by the following probability density function [38]:

$$f(\mathbf{x}) = \sum_{j=0}^{k-1} p_j \Phi(\mathbf{x}|\boldsymbol{\mu}_j, \Sigma_j), \tag{13.1}$$

where $\boldsymbol{\mu}_j$ is the mean of the jth component, Σ_j is the variance of the jth component, p_j is the mixing proportion of the jth component, and Φ is the probability density function of the multivariate Gaussian distribution.

DOI: 10.1201/9781003592648-13

The mixing proportions $p_0, p_1, \cdots, p_{k-1}$ in Equation (13.1) satisfy the following conditions:

(a) $p_j \in (0, 1)$ for $j = 0, 1, \cdots, k - 1$;

(b) The sum of the mixing proportions is equal to 1, i.e.,

$$\sum_{j=0}^{k-1} p_j = 1.$$

The probability density function of the multivariate Gaussian distribution is defined as

$$\Phi(\mathbf{x}|\boldsymbol{\mu}, \Sigma) = \frac{1}{\sqrt{(2\pi)^d |\Sigma|}} \exp\left(-\frac{1}{2}(\mathbf{x} - \boldsymbol{\mu})^T \Sigma^{-1}(\mathbf{x} - \boldsymbol{\mu})\right), \qquad (13.2)$$

where $\boldsymbol{\mu}$ is the mean, Σ is the variance matrix, and $|\Sigma|$ is the determinant of Σ. Here we assume that $\mathbf{x}$ and $\boldsymbol{\mu}$ are column vectors.

There are two approaches to cluster a dataset based on the Gaussian mixture model [38]: the mixture approach and the classification approach. In the mixture approach, the likelihood is maximized over the mixture parameters (i.e., $\boldsymbol{\mu}_j$ and Σ_j). In the classification approach, the likelihood is maximized over the mixture parameters as well as over the identifying labels of the mixture component origin for each record.

In this chapter, we implement the Gaussian mixture model-based clustering algorithm based on the first approach, i.e., the mixture approach. In this approach, the parameters that need to be estimated are

$$\Theta = (p_0, p_1, \cdots, p_{k-1}, \boldsymbol{\mu}_0, \boldsymbol{\mu}_1, \cdots, \boldsymbol{\mu}_{k-1}, \Sigma_0, \Sigma_1, \cdots, \Sigma_{k-1}).$$

We use the EM algorithm [178] to estimate these parameters by maximizing the log-likelihood. Given the dataset X, the log-likelihood is defined as

$$\mathcal{L}(\Theta; X) = \sum_{i=0}^{n-1} \ln\left(\sum_{j=0}^{k-1} p_j \Phi(\mathbf{x}_i|\boldsymbol{\mu}_j, \Sigma_j)\right). \qquad (13.3)$$

To estimate the parameter Θ, the EM algorithm starts with an initial parameter $\Theta^{(0)}$ and repeats the E-step and the M-step until it converges or the maximum number of iterations is reached. In the E-step, the conditional probabilities $t_j(\mathbf{x}_i)$ $(0 \le j \le k - 1, 0 \le i \le n - 1)$ that $\mathbf{x}_i$ comes from the jth component are calculated according to the following equation [38]:

$$t_j(\mathbf{x}_i) = \frac{p_j \Phi(\mathbf{x}_i|\boldsymbol{\mu}_j, \Sigma_j)}{\sum\limits_{s=0}^{k-1} p_s \Phi(\mathbf{x}_i|\boldsymbol{\mu}_s, \Sigma_s)}, \qquad (13.4)$$

where $p_j, \boldsymbol{\mu}_j, \Sigma_j$ $(0 \le j \le k - 1)$ are current estimates of Θ.

In the M-step, the parameter Θ is estimated based on the conditional probabilities $t_j(\mathbf{x}_i)$ $(0 \leq j \leq k-1, 0 \leq i \leq n-1)$ according to the following equations [38]:

$$p_j = \frac{\sum\limits_{i=0}^{n-1} t_j(\mathbf{x}_i)}{\sum\limits_{s=0}^{k-1} \sum\limits_{i=0}^{n-1} t_s(\mathbf{x}_i)}, \tag{13.5a}$$

$$\boldsymbol{\mu}_j = \frac{\sum\limits_{i=0}^{n-1} t_j(\mathbf{x}_i)\mathbf{x}_i}{\sum\limits_{i=0}^{n-1} t_j(\mathbf{x}_i)}, \tag{13.5b}$$

$$
\begin{aligned}
\Sigma_j &= \frac{\sum\limits_{i=0}^{n-1} t_j(\mathbf{x}_i)(\mathbf{x}_i - \boldsymbol{\mu}_j) \cdot (\mathbf{x}_i - \boldsymbol{\mu}_j)^T}{\sum\limits_{i=0}^{n-1} t_j(\mathbf{x}_i)} \\
&= \frac{1}{\sum\limits_{i=0}^{n-1} t_j(\mathbf{x}_i)} \begin{pmatrix} t_j(\mathbf{x}_1)(\mathbf{x}_1 - \boldsymbol{\mu}_j)^T \\ t_j(\mathbf{x}_2)(\mathbf{x}_2 - \boldsymbol{\mu}_j)^T \\ \vdots \\ t_j(\mathbf{x}_n)(\mathbf{x}_n - \boldsymbol{\mu}_j)^T \end{pmatrix}^T \begin{pmatrix} (\mathbf{x}_1 - \boldsymbol{\mu}_j)^T \\ (\mathbf{x}_2 - \boldsymbol{\mu}_j)^T \\ \vdots \\ (\mathbf{x}_n - \boldsymbol{\mu}_j)^T \end{pmatrix},
\end{aligned}
\tag{13.5c}
$$

for $j = 0, 1, \cdots, k-1$.

In our implementation, the initial parameter $\Theta^{(0)}$ are chosen as follows:

$$p_j = \frac{1}{k}, \tag{13.6a}$$

$$\boldsymbol{\mu}_j = \mathbf{x}_{i_j}, \tag{13.6b}$$

$$\Sigma_j = \text{Diag}(\sigma_0^2, \sigma_1^2, \cdots, \sigma_{d-1}^2), \tag{13.6c}$$

for $j = 0, 1, \cdots, k-1$, where i_j $(0 \leq j \leq k-1)$ are random integers chose from $\{0, 1, \cdots, n-1\}$ and

$$\sigma_s^2 = \frac{1}{n-1} \sum_{i=0}^{n-1} x_{is}^2 - \frac{1}{n(n-1)} \left(\sum_{i=0}^{n-1} x_{is} \right)^2, \quad s = 0, 1, \cdots, d-1.$$

Here x_{is} is the sth component of $\mathbf{x}_i$ and σ_s^2 is the sample variance of the sth attribute.

Once the parameter Θ is determined by the EM algorithm, the cluster memberships $\gamma_0, \gamma_1, \cdots, \gamma_{n-1}$ can be derived from Equation (13.4) as follows:

$$\gamma_i = \operatorname*{argmax}_{0 \leq j \leq k-1} t_j(\mathbf{x}_i), \quad i = 0, 1, \cdots, n-1. \tag{13.7}$$

13.2   Implementation

In this section, we implement the Gaussian mixture clustering algorithm in Python with the help of the NumPy and SciPy libraries. Like the k-means algorithm, the Gaussian mixture clustering algorithm can be implemented as follows:

```
def gmc(X, k=3, tol=1e-8, maxit=100):
    X = np.ascontiguousarray(X)
    n, d = X.shape
    mixingProp = np.ones(k) / k
    ind = np.random.choice(n, k, replace=False)
    clusterCenters = X[ind,:]
    varianceMatrices = [np.diag(np.var(X, axis=0))] * k
    density = np.zeros((n, k))
    for i in range(k):
        density[:,i] = mixingProp[i]*multivariate_normal(
            mean=clusterCenters[i,:], cov=varianceMatrices[
            i]).pdf(X)
    conditionalProb = density / np.sum(density, axis=1,
        keepdims=True)
    logLikelihood = np.sum(np.log(np.sum(density, axis=1)))
        .item()
    numIter = 1
    while numIter < maxit:
        # M-step
        dv = np.sum(conditionalProb, axis=0)
        mixingProp = dv / np.sum(dv)
        for i in range(k):
            clusterCenters[i,:] = np.sum(X*conditionalProb
                [:,i].reshape((n,1)), axis=0)/np.sum(
                conditionalProb[:,i])
            Xc = X - clusterCenters[i,:]
            varianceMatrices[i] = np.matmul(np.transpose(Xc
                ), Xc * conditionalProb[:,i].reshape((n,1))
                )/np.sum(conditionalProb[:,i])
        # E-step
        density = np.zeros((n, k))
        for i in range(k):
            density[:,i] = mixingProp[i]*
                multivariate_normal(mean=clusterCenters[i
                ,:], cov=varianceMatrices[i]).pdf(X)
        conditionalProb = density / np.sum(density, axis=1,
            keepdims=True)
        logLikelihood_ = logLikelihood
        logLikelihood = np.sum(np.log(np.sum(density, axis
            =1))).item()
```

```
29      numIter += 1
30      if np.abs(logLikelihood - logLikelihood_) < tol:
31          break;
32  return conditionalProb, clusterCenters, mixingProp,
        logLikelihood, numIter
```

In Lines 4–7, we initialize the parameters, which include the mixture proportions, the cluster centers, and the covariance matrices. In Lines 8–12, we calculate the conditional probabilities and the log-likelihood function value based on the initial values of the parameters.

In the iterative process, we repeat the M-step and the E-step alternatively to update the parameters. The iterative process is terminated when the absolute difference of the log-likelihood function values from two consecutive iterations is less than the tolerance.

In our implementation, we use the `multivariate_normal` from the SciPy library to calculate the probability density function of the multivariate normal distribution. We try to avoid loops as much as possible to improve the speed of the program.

The following Python function implements the multiple run of the Gaussian mixture clustering algorithm:

```
1  def gmc2(X, k=3, numrun=10, maxit=100):
2      bestCP, bestCC, bestMP, bestLL, bestIters = gmc(X, k,
           maxit)
3      vLL = np.zeros(numrun)
4      vLL[0] = bestLL
5      for i in range(numrun-1):
6          cp, cc, mp, ll, iters = gmc(X, k, maxit)
7          vLL[i+1] =ll
8          if ll > bestLL:
9              bestCP, bestCC, bestMP, bestLL, bestIters = cp,
                  cc, mp, ll, iters
10     return bestCP, bestCC, bestMP, bestLL, bestIters, vLL
```

Unlike the k-means algorithm, the Gaussian mixture clustering algorithm tries to maximize the log-likelihood function, which serves as the objective function. Hence, the best run is determined to be the run with the maximum log-likelihood function value.

13.3   Examples

In this section, we apply the Gaussian mixture clustering algorithm implemented in the previous section to a synthetic dataset and the Iris dataset. Before we present the examples, we need to load the necessary libraries by running the following code:

```
import numpy as np
import matplotlib.pyplot as plt
from scipy.stats import multivariate_normal
from sklearn.datasets import make_blobs
from ucimlrepo import fetch_ucirepo
from dcutil import createCM
```

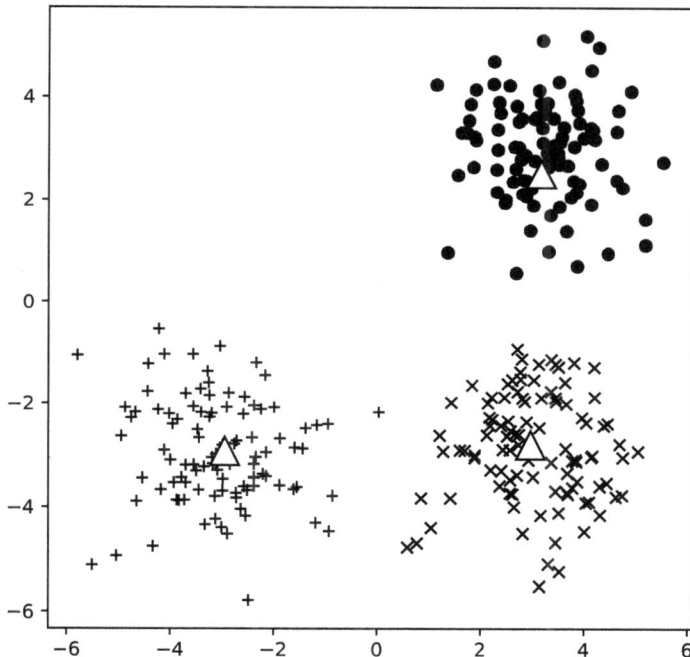

FIGURE 13.1: Clusters produced by the Gaussian mixture clustering algorithm on the synthetic dataset. Cluster centers are indicated by the triangles.

First, let us illustrate the Gaussian mixture clustering algorithm with a synthetic dataset. To do that, we use the following piece of code:

```
centers = [[3, 3], [-3, -3], [3, -3]]
X, y = make_blobs(n_samples=300, centers=centers,
    cluster_std=1, random_state=1)

bcp, bcc, bmp, bll, biters, vLL = gmc2(X, k=3, numrun=100)
yhat = np.argmax(bcp, axis=1)
cm1 = createCM(y, yhat)
print(cm1)
print([bll, biters])
```

In the above code, we run the Gaussian mixture clustering algorithm 100 times. Executing the above code, we see the following output:

```
       0    1    2
0   0  100    0
1   0    0  100
2  99    1    0
[-1210.3400278163012, 3]
```

From the resulting confusion matrix, we see that only one record was misclassified. The best log-likelihood function value was about −1210.34 and the best run terminated in three iterations. Figure 13.1 shows the three clusters obtained by the best run of the Gaussian mixture clustering algorithm on the synthetic dataset.

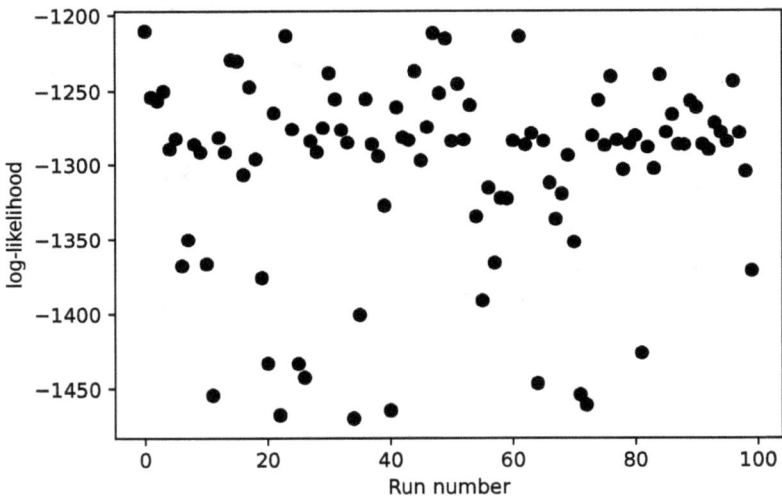

FIGURE 13.2: Log-likelihood function values of 100 runs of the Gaussian mixture clustering algorithm on the synthetic dataset.

We can plot the log-likelihood function values from the 100 runs as follows:

```
fig, ax = plt.subplots(1, 1, figsize=(6, 4))
ax.scatter(range(len(vLL)), vLL, color="black")
ax.set_xlabel("Run number")
ax.set_ylabel("log-likelihood")
fig.savefig("3001l.pdf", bbox_inches='tight')
```

Figure 13.2 shows the resulting plot. From the figure, we see that the log-likelihood function values from the 100 runs vary a lot. This indicates that the Gaussian mixture clustering algorithm is sensitive to the initial cluster centers.

To apply the Gaussian mixture clustering algorithm to the Iris dataset 100 times, we use the following code:

```
iris = fetch_ucirepo(id=53)
X = iris.data.features
y = iris.data.targets

bcp, bcc, bmp, bll, biters, vLL = gmc2(X, k=3, numrun=100)
yhat = np.argmax(bcp, axis=1)
cm1 = createCM(y, yhat)
print(cm1)
print([bll, biters])
```

Executing the above block of code gives the following output:

```
                   0    1    2
Iris-setosa       50    0    0
Iris-versicolor    0    3   47
Iris-virginica     0   44    6
[-204.81921785043306, 3]
```

From the output, we see that 9 records were misclassified in the best run, which was terminated after three iterations.

Figure 13.3 shows the log-likelihood function values of the 100 runs of the Gaussian mixture clustering algorithm on the Iris dataset. The log-likelihood

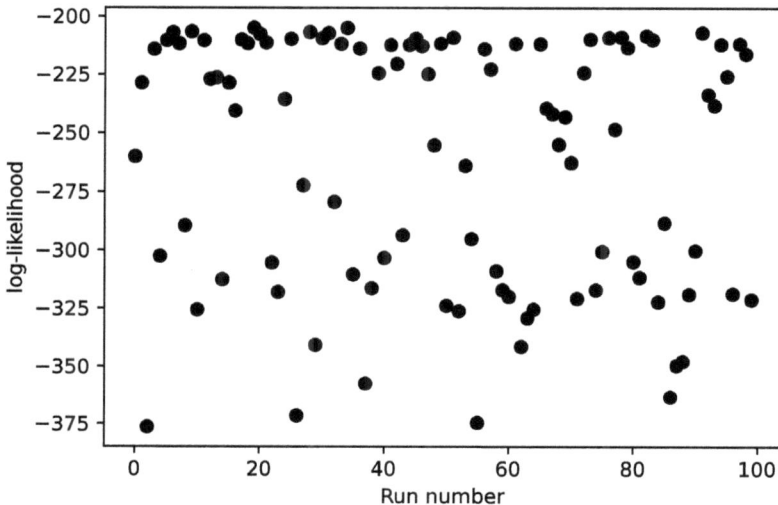

FIGURE 13.3: Log-likelihood function values of 100 runs of the Gaussian mixture clustering algorithm on the Iris dataset.

function values have a big range, indicating that the Gaussian mixture clustering algorithm is sensitive to the initial cluster centers.

13.4 Summary

In this chapter, we implemented one of the Gaussian mixture clustering algorithms summarized in [38]. The Gaussian mixture clustering algorithms are model-based clustering algorithms. Mode-based clustering is a major approach to cluster analysis and has a long history. [27] presented a survey of cluster analysis based on probabilistic models. Recent work on model-based clustering can be found in [276].

14

The KMTD Algorithm

The KMTD (k-means-type clustering based on the t-distribution) algorithm [258] is a model-based clustering algorithm developed for clustering noisy data. In this chapter, we introduce the KMTD algorithm and its implementation.

14.1 Description of the Algorithm

The KMTD algorithm is motivated by how the k-means algorithm is derived a Gaussian mixture model [258]. In the KMTD algorithm, the data points are assumed to be drawn from a special multivariate t-mixture model. Since the t-distribution has heavier tails than the Gaussian distribution, the KMTD algorithm is expected to be more robust than Gaussian mixture clustering for handling noisy data.

To describe the t-mixture model (TMM), we let $f(\mathbf{x}; \nu, \boldsymbol{\mu}, \boldsymbol{\Sigma})$ be the probability density function of a multivariate t-distribution with parameters ν, $\boldsymbol{\mu}$, $\boldsymbol{\Sigma}$. The probability density function is defined as [101]:

$$f(\mathbf{x}; \nu, \boldsymbol{\mu}, \boldsymbol{\Sigma}) = \frac{\Gamma\left(\frac{\nu+d}{2}\right)\left[1 + \frac{1}{\nu}(\mathbf{x} - \boldsymbol{\mu})^T \boldsymbol{\Sigma}^{-1}(\mathbf{x} - \boldsymbol{\mu})\right]^{-\frac{\nu+d}{2}}}{\Gamma\left(\frac{\nu}{2}\right)\nu^{\frac{d}{2}}\pi^{\frac{d}{2}}|\boldsymbol{\Sigma}|^{\frac{1}{2}}}, \tag{14.1}$$

where $\Gamma(\cdot)$ is the gamma function and d denotes the number of dimensions of the data. It is worth pointing out that the covariance matrix of the t-distribution is not $\boldsymbol{\Sigma}$ but $\frac{\nu}{\nu-2}\boldsymbol{\Sigma}$, which exists only when $\nu > 2$.

The probability density function of a mixture of multivariate t-distributions can be expressed as:

$$f(\mathbf{x}) = \sum_{l=0}^{k-1} p_l f(\mathbf{x}; \nu_l, \boldsymbol{\mu}_l, \boldsymbol{\Sigma}_l), \tag{14.2}$$

where k is the number of components, p_l's are positive coefficients that satisfy the following condition

$$\sum_{l=0}^{k-1} p_l = 1. \tag{14.3}$$

DOI: 10.1201/9781003592648-14

183

The KMTD algorithm is derived by assuming a special case of the TMM where the mixture components are spherical and share the same parameters $\boldsymbol{\Sigma}$, i.e., $\boldsymbol{\Sigma}_l = \sigma^2 \mathbf{I}$ and $\nu_l = \nu$ for $l = 0, 1, \ldots, k-1$, where $\mathbf{I}$ is a $d \times d$ identity matrix, $\sigma > 0$, and $\nu > 0$. The probability density function of the special t-distribution is

$$
f(\mathbf{x}; \nu, \boldsymbol{\mu}, \sigma) = \frac{\Gamma\left(\frac{\nu+d}{2}\right) \left[1 + \frac{1}{\nu\sigma^2}\|\mathbf{x} - \boldsymbol{\mu}\|^2\right]^{-\frac{\nu+d}{2}}}{\Gamma\left(\frac{\nu}{2}\right) \nu^{\frac{d}{2}} \pi^{\frac{d}{2}} \sigma^d}, \tag{14.4}
$$

where $\| \cdot \|$ is the L_2 norm.

The special TMM has the following parameters:

$$
\Theta = (p_0, p_1, \cdots, p_{k-1}, \boldsymbol{\mu}_0, \boldsymbol{\mu}_1, \cdots, \boldsymbol{\mu}_{k-1}).
$$

The parameters ν and σ of the t-distribution are user-specified. Similar to the Gaussian mixture clustering algorithm, we use the EM algorithm for parameter estimation.

Let X be the random variable that represents the data and let Z be the latent variable that represents the cluster membership of X. That is, $Z = j$ if X is generated by the jth component, where $0 \leq j \leq k-1$. Let $D = \{\mathbf{x}_0, \mathbf{x}_1, \ldots, \mathbf{x}_{n-1}\}$ be a set of n observations.

In the E-step, we calculate the expected value of the log-likelihood function of Θ with respect to the current conditional distribution of Z given the data D and the current estimates of the parameters Θ'. This is done as follows:

$$
\begin{aligned}
\mathcal{L}(\Theta|D, \Theta') &= E\left[\ln \prod_{i=0}^{n-1} f(\mathbf{x}_i; \nu, \boldsymbol{\mu}, \sigma^2 \mathbf{I}) \,\middle|\, \Theta'\right] \\
&= E\left[\sum_{i=0}^{n-1} \ln f(\mathbf{x}_i; \nu, \boldsymbol{\mu}, \sigma^2 \mathbf{I}) \,\middle|\, \Theta'\right] \\
&= E\left[\sum_{i=0}^{n-1} \ln \left(\sum_{j=0}^{k-1} P(Z = j) f(\mathbf{x}_i; \nu_j, \boldsymbol{\mu}_j, \sigma^2 \mathbf{I})\right) \,\middle|\, \Theta'\right] \\
&= \sum_{i=0}^{n-1} \ln \left(\sum_{j=0}^{k-1} p_j f(\mathbf{x}_i; \nu, \boldsymbol{\mu}_j, \sigma)\right), \tag{14.5}
\end{aligned}
$$

which is just the log-likelihood. Here $p_j = P(Z = j)$ for $j = 0, 1, \ldots, k-1$.

The conditional probabilities of assignments given the current estimates of the parameters are calculated as follows. By using the Bayes' rule, we can calculate the conditional probability of Z_i as follows:

$$
\begin{aligned}
P(Z_i = l | X_i = \mathbf{x}_i, \Theta) &= \frac{P(Z_i = l, X_i = \mathbf{x}_i | \Theta)}{P(X_i = \mathbf{x}_i | \Theta)} \\
&= \frac{P(X_i = \mathbf{x}_i | Z_i = l, \Theta) P(Z_i = l)}{P(X_i = \mathbf{x}_i | \Theta)} \\
&= \frac{f(\mathbf{x}_i; \nu_l, \boldsymbol{\mu}_l, \boldsymbol{\Sigma}_l) p_l}{\sum_{h=1}^{k} p_h f(\mathbf{x}_i; \nu_h, \boldsymbol{\mu}_h, \boldsymbol{\Sigma}_h)}.
\end{aligned}
\tag{14.6}
$$

Combining Equation (14.6) and Equation (14.4), we can compute the conditional probabilities as follows:

$$
\begin{aligned}
\gamma_{i,l} = P(Z_i = l | X_i = \mathbf{x}_i) &= \frac{p_l \left[1 + \frac{1}{\nu\sigma^2} \|\mathbf{x}_i - \boldsymbol{\mu}_l\|^2\right]^{-\frac{\nu+d}{2}}}{\sum_{h=0}^{k-1} p_h \left[1 + \frac{1}{\nu\sigma^2} \|\mathbf{x}_i - \boldsymbol{\mu}_h\|^2\right]^{-\frac{\nu+d}{2}}} \\
&= \frac{p_l \left[\nu\sigma^2 + \|\mathbf{x}_i - \boldsymbol{\mu}_l\|^2\right]^{-\frac{\nu+d}{2}}}{\sum_{h=0}^{k-1} p_h \left[\nu\sigma^2 + \|\mathbf{x}_i - \boldsymbol{\mu}_h\|^2\right]^{-\frac{\nu+d}{2}}}.
\end{aligned}
\tag{14.7}
$$

In the M-step, we maximize the log-likelihood function to estimate the parameters by fixing $\gamma_{i,l}$'s. Note that the weighting coefficients are constrained by the conditions given in Equation (14.3). Using the method of Lagrange multipliers, we maximize the following objective function

$$
\begin{aligned}
L_2(\boldsymbol{\mu}_0, \ldots, &\boldsymbol{\mu}_{k-1}, p_0, \ldots, p_{k-1}, \lambda) \\
&= \sum_{i=0}^{n-1} \ln \left(\sum_{j=0}^{k-1} p_j f(\mathbf{x}_i; \nu, \boldsymbol{\mu}_j, \sigma)\right) - \lambda \left(\sum_{l=0}^{k-1} p_l - 1\right).
\end{aligned}
\tag{14.8}
$$

Equating the derivatives of L_2 with respect to p_l's and λ to zero, we get

$$
\frac{\partial L_2}{\partial p_l} = \sum_{i=0}^{n-1} \frac{f(\mathbf{x}_i; \nu, \boldsymbol{\mu}_l, \sigma)}{\sum_{h=1}^{k} p_h f(\mathbf{x}_i; \nu, \boldsymbol{\mu}_h, \sigma)} - \lambda = 0.
$$

Plugging Equation (14.7) into the above equation gives

$$
\sum_{i=0}^{n-1} \frac{\gamma_{i,l}}{p_l} - \lambda = 0.
\tag{14.9}
$$

Combining Equation (14.9) and Equation (14.3), we get

$$
p_l = \frac{1}{n} \sum_{i=0}^{n-1} \gamma_{i,l}, \quad l = 0, 1, 2, \ldots, k - 1.
\tag{14.10}
$$

Similarly, equating the derivatives of L_2 with respect to $\boldsymbol{\mu}_l$'s to zero, we get

$$\frac{\partial L_2}{\partial \boldsymbol{\mu}_l} = \sum_{i=0}^{n-1} \frac{p_l f(\mathbf{x}_i; \nu, \boldsymbol{\mu}_l, \sigma) \dfrac{-\frac{2}{\nu\sigma^2}(\mathbf{x}_i - \boldsymbol{\mu}_l)^T}{1 + \frac{1}{\nu\sigma^2}\|\mathbf{x}_i - \boldsymbol{\mu}_l\|^2}}{\sum_{h=0}^{k-1} p_h f(\mathbf{x}_i; \nu, \boldsymbol{\mu}_l, \sigma)} = 0.$$

Combining Equation (14.9) and the above equation gives

$$\sum_{i=0}^{n-1} \gamma_{i,l} \frac{-\frac{2}{\nu\sigma^2}(\mathbf{x}_i - \boldsymbol{\mu}_l)^T}{1 + \frac{1}{\nu\sigma^2}\|\mathbf{x}_i - \boldsymbol{\mu}_l\|^2} = 0$$

or

$$\boldsymbol{\mu}_l = \frac{\sum_{i=0}^{n-1} \dfrac{\gamma_{i,l}\mathbf{x}_i}{\nu\sigma^2 + \|\mathbf{x}_i - \boldsymbol{\mu}_l\|^2}}{\sum_{i=0}^{n-1} \dfrac{\gamma_{i,l}}{\nu\sigma^2 + \|\mathbf{x}_i - \boldsymbol{\mu}_l\|^2}}. \tag{14.11}$$

Although Equation (14.11) does not give an explicit solution of $\boldsymbol{\mu}_l$, it gives a recursive formula to solve $\boldsymbol{\mu}_l$.

In the KMTD algorithm, the parameters ν and σ are specified by the user. Those two parameters are not estimated from the data during the clustering process.

14.2   Implementation

The implementation of the KMTD algorithm is similar to that of the Gaussian mixture clustering algorithm (see Section 13.2). The following function implements the KMTD algorithm:

```
1  def kmtd(X, k=3, nu=3, sigma=None, tol=1e-8, maxit=100):
2      X = np.ascontiguousarray(X)
3      n, d = X.shape
4      mixingProp = np.ones(k) / k
5      ind = np.random.choice(n, k, replace=False)
6      clusterCenters = X[ind,:]
7      if sigma is None:
8          sigma = np.sqrt(np.mean(np.var(X, axis=0)))*(nu-2)/
              nu)
9      density = np.zeros((n, k))
10     for i in range(k):
11         density[:,i] = mixingProp[i]*multivariate_t(loc=
              clusterCenters[i,:], shape=sigma**2*np.eye(d),
              df=nu).pdf(X)
```

```
12    conditionalProb = density / np.sum(density, axis=1,
         keepdims=True)
13    logLikelihood = np.sum(np.log(np.sum(density, axis=1)))
         .item()
14    dm = np.zeros((n,k))
15    numIter = 1
16    while numIter < maxit:
17        # M-step
18        dv = np.sum(conditionalProb, axis=0)
19        mixingProp = dv / np.sum(dv)
20        for i in range(k):
21            dm[:,i] = np.sum(np.square(X-clusterCenters[i
                 ,:]), axis=1)
22        for i in range(k):
23            vw = conditionalProb[:,i]/(nu*sigma**2 + dm[:,i
                 ])
24            clusterCenters[i,:] = np.sum(X*vw.reshape((n,1)
                 ), axis=0)/np.sum(vw)
25        # E-step
26        density = np.zeros((n, k))
27        for i in range(k):
28            density[:,i] = mixingProp[i]*multivariate_t(loc
                 =clusterCenters[i,:], shape=sigma**2*np.eye
                 (d), df=nu).pdf(X)
29        conditionalProb = density / np.sum(density, axis=1,
             keepdims=True)
30        logLikelihood_ = logLikelihood
31        logLikelihood = np.sum(np.log(np.sum(density, axis
             =1))).item()
32        numIter += 1
33        if np.abs(logLikelihood - logLikelihood_) < tol:
34            break;
35    return conditionalProb, clusterCenters, mixingProp,
         logLikelihood, numIter
```

In the above function, the default value of the parameter ν is 3. The parameter σ will be estimated from the data if it is not supplied by the user.

The following function implements the multiple run of the KMTD algorithm:

```
1  def kmtd2(X, k=3, nu=3, sigma=None, numrun=10, maxit=100):
2      bestCP, bestCC, bestMP, bestLL, bestIters = kmtd(X, k=k
           , nu=nu, sigma=sigma, maxit=maxit)
3      vLL = np.zeros(numrun)
4      vLL[0] = bestLL
5      for i in range(numrun-1):
6          cp, cc, mp, ll, iters = kmtd(X, k=k, nu=nu, sigma=
               sigma, maxit=maxit)
7          vLL[i+1] =ll
```

```
 8    if ll > bestLL:
 9        bestCP, bestCC, bestMP, bestLL, bestIters = cp,
             cc, mp, ll, iters
10    return bestCP, bestCC, bestMP, bestLL, bestIters, vLL
```

Since we want to maximize the log-likelihood function, the best run is determined to be the run that has the maximum log-likelihood function value.

14.3   Examples

In this section, we apply the KMTD algorithm implemented in the previous
section to a synthetic dataset and the Iris dataset. Before we run the algorithm,
we need to load the necessary libraries by running the following code:

```
1  import numpy as np
2  import matplotlib.pyplot as plt
3  from scipy.stats import multivariate_t
4  from sklearn.datasets import make_blobs
5  from ucimlrepo import fetch_ucirepo
6  from dcutil import createCM
```

First, let us illustrate the KMTD algorithm with a synthetic dataset. To
do that, we use the following piece of code:

```
1  centers = [[3, 3], [-3, -3], [3, -3]]
2  X, y = make_blobs(n_samples=300, centers=centers,
       cluster_std=1, random_state=1)
3
4  bcp, bcc, bmp, bll, biters, vLL = kmtd2(X, k=3, nu=3,
       numrun=100)
5  yhat = np.argmax(bcp, axis=1)
6  cm1 = createCM(y, yhat)
7  print(cm1)
8  print([bll, biters])
```

In the above code, we run the KMTD algorithm 100 times. Executing the
code above produces the following output:

```
1       0    1    2
2  0    0    0   100
3  1    1   99    0
4  2   99    0    1
5  [-1330.0969096201227, 15]
```

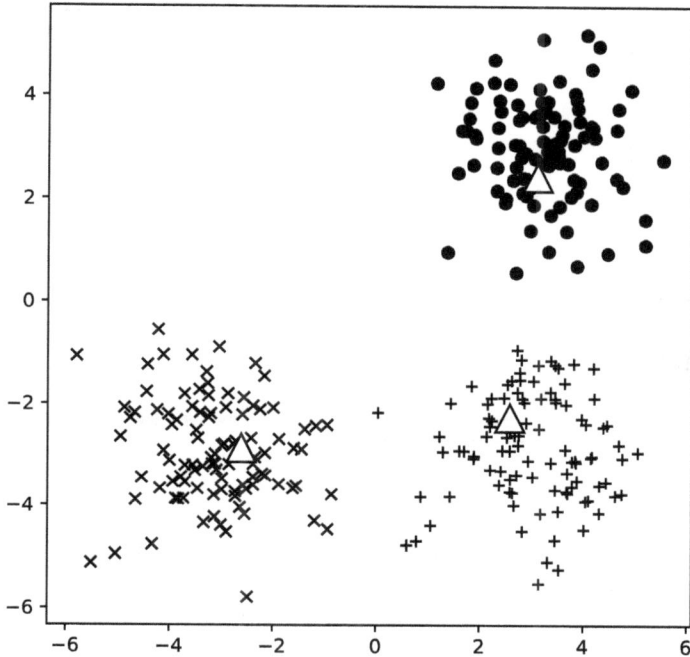

FIGURE 14.1: Clusters produced by the KMTD algorithm on the synthetic dataset. Cluster centers are indicated by the triangles.

From the resulting confusion matrix, we see that two records were misclassified. The best log-likelihood function value was about -1330.1 and the best run terminated in 15 iterations. Figure 14.1 shows the three clusters obtained by the best run of the KMTD algorithm on the synthetic dataset.

We can plot the log-likelihood function values from the 100 runs as follows:

```
fig, ax = plt.subplots(1, 1, figsize=(6, 4))
ax.scatter(range(len(vLL)), vLL, color="black")
ax.set_xlabel("Run number")
ax.set_ylabel("log-likelihood")
fig.savefig("300llkmtd.pdf", bbox_inches='tight')
```

Figure 14.2 shows the resulting plot. From the figure, we see that the log-likelihood function values from the 100 runs form two clear groups. This indicates that the KMTD algorithm is not sensitive to the initial cluster centers.

To apply the KMTD algorithm to the Iris dataset 100 times, we use the following code:

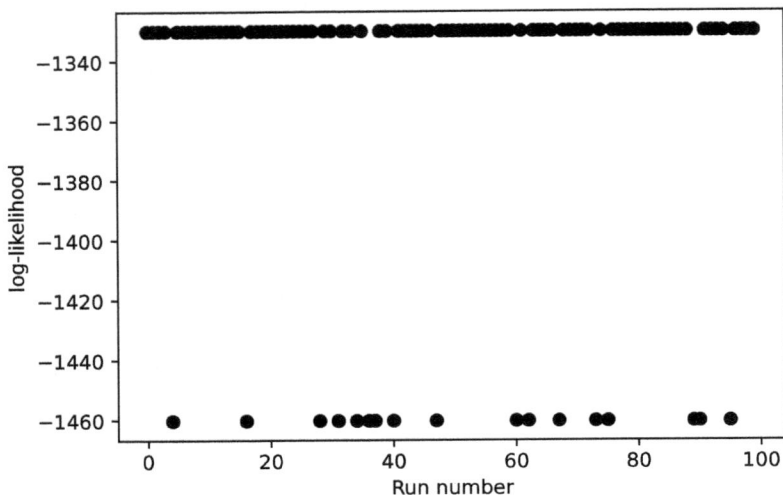

FIGURE 14.2: Log-likelihood function values of 100 runs of the KMTD algorithm on the synthetic dataset.

```
1  iris = fetch_ucirepo(id=53)
2  X = iris.data.features
3  y = iris.data.targets
4
5  bcp, bcc, bmp, bll, biters, vLL = kmtd2(X, k=3, nu=3,
      numrun=100)
6  yhat = np.argmax(bcp, axis=1)
7  cm1 = createCM(y, yhat)
8  print(cm1)
9  print([bll, biters])
```

Executing the above block of code gives the following output:

```
1                  0    1    2
2  Iris-setosa     50   0    0
3  Iris-versicolor 0    48   2
4  Iris-virginica  0    14   36
5  [-512.6709994811413, 69]
```

From the output, we see that 16 records were misclassified in the best run, which was terminated after 69 iterations.

Figure 14.3 shows the log-likelihood function values of the 100 runs of the KMTD algorithm on the Iris dataset. The log-likelihood function values also form two clear groups, indicating that the KMTD algorithm is not sensitive to the initial cluster centers.

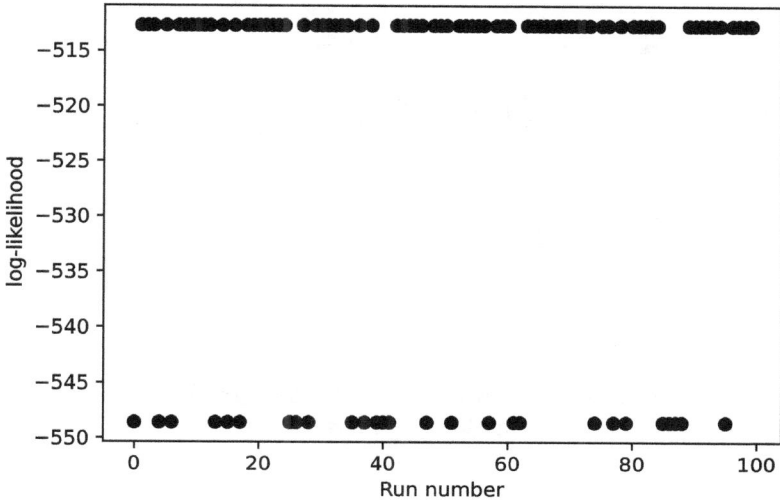

FIGURE 14.3: Log-likelihood function values of 100 runs of the KMTD algorithm on the Iris dataset.

14.4 Summary

In this chapter, we implemented the KMTD (k-means-type clustering based on the t-distribution) algorithm [258]. The KMTD algorithm is a model-based clustering algorithm developed to clusterize noisy data. It is derived under the assumption of a special case of the t-mixture model where the mixture components are spherical and share the same parameters. Experimental results suggest that the KMTD algorithm is less sensitive to initial cluster centers compared to the Gaussian mixture model.

15

The Probability Propagation Algorithm

The probability propagation algorithm [98] is a graph-based clustering algorithm that can identify arbitrarily shaped clusters. In this chapter, we introduce this algorithm and its implementations.

15.1 Description of the Algorithm

The probability propagation (PP) algorithm consists of two steps [98]: first, create a stochastic matrix based on a bandwidth parameter; second, propagate the probabilities by raising the stochastic matrix to powers.

To describe the PP algorithm, we let $D = \{\mathbf{x}_0, \mathbf{x}_1, \ldots, \mathbf{x}_{n-1}\}$ denote a data set containing n records. Let W denote the stochastic matrix, which is an $n \times n$ matrix. The (i, j)-th entry of W represents the probability that $\mathbf{x}_i$ chooses $\mathbf{x}_j$ as its attractor. Here an attractor can be thought as a cluster center. The stochastic matrix W has the following property:

$$\sum_{j=0}^{n-1} w_{ij} = 1, \quad i = 0, 1, \ldots, n-1, \tag{15.1}$$

where w_{ij} denotes the (i, j)-th entry of W. The above equations hold because each record can choose itself as its attractor.

The stochastic matrix is initialized in such a way that the probability of a record selecting a neighbor as its attractor is proportional to the local density of the neighbor. Mathematically, the stochastic matrix is initialized as follows. The local density of a record $\mathbf{y}$ is defined as

$$V(\mathbf{y}) = \sum_{\mathbf{x} \in N(\mathbf{y})} K\left(\frac{d(\mathbf{x}, \mathbf{y})}{\delta}\right), \tag{15.2}$$

where δ is the bandwidth parameter, $d(\cdot, \cdot)$ is a distance function, $K(\cdot)$ is a kernel function, and $N(\mathbf{y})$ denotes the set of neighbors of $\mathbf{y}$. The set of

DOI: 10.1201/9781003592648-15

neighbors is defined as follows:

$$N(\mathbf{y}) = \{\mathbf{x} \in D : d(\mathbf{x}, \mathbf{y}) < \delta\}.$$

Once we have determined the sets of neighbors for all records. We define the following matrix W':

$$w'_{ij} = \begin{cases} V(\mathbf{x}_j), & \text{if } \mathbf{x}_j \in N(\mathbf{x}_i), \\ 0, & \text{if } \mathbf{x}_j \notin N(\mathbf{x}_i). \end{cases}$$

Then for each $i = 0, 1, \ldots, n - 1$, we find a permutation $(i_0, i_1, \ldots, i_{n-1})$ of $(0, 1, \ldots, n - 1)$ such that

$$w'_{i,i_0} \geq w'_{i,i_1} \geq \cdots \geq w'_{i,i_{n-1}}.$$

Then the initial stochastic matrix is formulated as follows:

$$w_{ij} = \begin{cases} \dfrac{w'_{ij}}{\sum_{r \in \{i_0, i_1, \ldots, i_s\}} w'_{ir}}, & \text{if } j \in \{i_0, i_1, \ldots, i_s\}, \\ 0, & \text{if } j \notin \{i_0, i_1, \ldots, i_s\}, \end{cases} \tag{15.3}$$

where s is a parameter used to control the shape of the clusters.

Once the stochastic matrix is initialized, the probabilities are propagated as follows:

$$w_{ij} \leftarrow \sum_{l=0}^{n-1} w_{il} w_{lj}, \quad i, j = 0, 1, \ldots, n - 1,$$

or in matrix form

$$W \leftarrow W^2. \tag{15.4}$$

The above probability propagation operation is repeated until the set of attractors does not change. For $i = 0, 1, \ldots, n - 1$, let $\mathbf{x}_{i^*}$ be the attractor of $\mathbf{x}_i$. The attractor's index can be identified as follows:

$$i^* = \arg \max_{0 \leq j \leq n-1} w^*_{ij},$$

where W^* is the converged stochastic matrix.

There are several choices of kernel functions and distance functions. For example, the kernel function can be the Gaussian kernel, which is defined by

$$K(u) = \frac{1}{\sqrt{2\pi}} \exp\left(-\frac{u^2}{2}\right).$$

The distance function can be the Euclidean distance function. The bandwidth parameter influences the neighborhood structure, as it determines which points are considered neighbors based on the chosen distance function.

15.2   Implementation

Implementing the PP algorithm is simple in Python. To optimize performance, the implementation utilizes vectorized operations. This can be done by using functions provided by the NumPy library.

The following function implements the PP algorithm:

```
def pp(X, delta=None, s=5, maxit=100):
    X = np.ascontiguousarray(X)
    n, d = X.shape
    W = np.zeros((n,n))
    dm = squareform(pdist(X))
    if delta is None:
        delta = np.percentile(np.mean(dm, axis=0), 10)
    V = np.exp(-np.square(dm/delta)/2)
    ind = np.argsort(V)[:, -1:-s-1:-1]
    val = np.take_along_axis(V, ind, axis=-1)
    np.put_along_axis(W, ind, val, axis=-1)
    W = W / np.sum(W, axis=1, keepdims=True)
    numIter = 1
    attractors = set(np.argmax(W, axis=1))
    while numIter < maxit:
        W = np.linalg.matrix_power(W, 2)
        attractors_ = attractors.copy()
        attractors = set(np.argmax(W, axis=1))
        if attractors == attractors_:
            break
    return list(attractors), W
```

The `pp` function has four parameters, which include the dataset, the bandwidth, the number of top values, and the maximum number of iteration. By default, the bandwidth is estimated from the distance matrix. It is estimated to be the 10th percentile of the mean distances from all columns.

In the above code, we use the `pdist` function from the SciPy library to calculate the distance matrix. We use the `argsort` function from the NumPy library to get the indices of the top s values from each row of the matrix containing local densities. The corresponding values are extracted by using the `take_along_axis` function from the NumPy library. Then those values are assigned to the stochastic matrix by using the `put_along_axis` function.

The iterative process is also simple. The `matrix_power` function is used to raise the stochastic matrix to power 2. The iterative process terminates once the attractor sets from two consecutive iterations remain unchanged.

Since the stochastic matrix contains lots of zeros, we can use a sparse matrix to represent the stochastic matrix. The following function implements the PP algorithm by using a sparse stochastic matrix:

```
1  def pps(X, delta=None, s=5, maxit=100):
2      X = np.ascontiguousarray(X)
3      n, d = X.shape
4      W = lil_array((n,n))
5      dm = squareform(pdist(X))
6      if delta is None:
7          delta = np.percentile(np.mean(dm, axis=0), 10)
8      V = np.exp(-np.square(dm/delta)/2)
9      ind = np.argsort(V)[:, -1:-s-1:-1]
10     val = np.take_along_axis(V, ind, axis=-1)
11     np.put_along_axis(W, ind, val, axis=-1)
12     W = W / np.sum(W, axis=1).reshape((n,1))
13     numIter = 1
14     attractors = set(np.argmax(W, axis=1))
15     while numIter < maxit:
16         W = matrix_power(W, 2)
17         attractors_ = attractors.copy()
18         attractors = set(np.argmax(W, axis=1))
19         if attractors == attractors_:
20             break
21     return list(attractors), W
```

The `pps` function is different from the `pp` function in only a few lines. Many of the NumPy functions work for the sparse matrix. We do not need to change the code. We change the `matrix_power` function from the NumPy library to the one from the SciPy library. The necessary libraries required by the `pp` function and the `pps` function are given in the next section.

15.3   Examples

In this section, we illustrate the PP algorithm with two synthetic datasets and the Iris dataset. The first synthetic dataset contains spherically shaped clusters. The second synthetic dataset contains chain-like clusters.

The following code shows the application of the PP algorithm to the first synthetic dataset:

```
1  centers = [[3, 3], [-3, -3], [3, -3]]
2  X, y = make_blobs(n_samples=300, centers=centers,
       cluster_std=1, random_state=1)
3
4  ci, W = pp(X)
5  yhat = np.argmax(W, axis=1)
6  cm1 = createCM(y, yhat)
7  print(cm1)
```

In the above code, we use the default values of the parameters. Executing the above block of code gives the following output:

```
     20    54   90
0     0   100    0
1    99     0    1
2     0     1   99
```

The confusion matrix shown above indicates that only two records were misclassified. The attractors are records with indices 20, 54, and 90.

Figure 15.1 shows the clusters obtained by applying the PP algorithm to the first synthetic dataset. From the plot, we see that the attractors are not exactly located at the centers of the clusters.

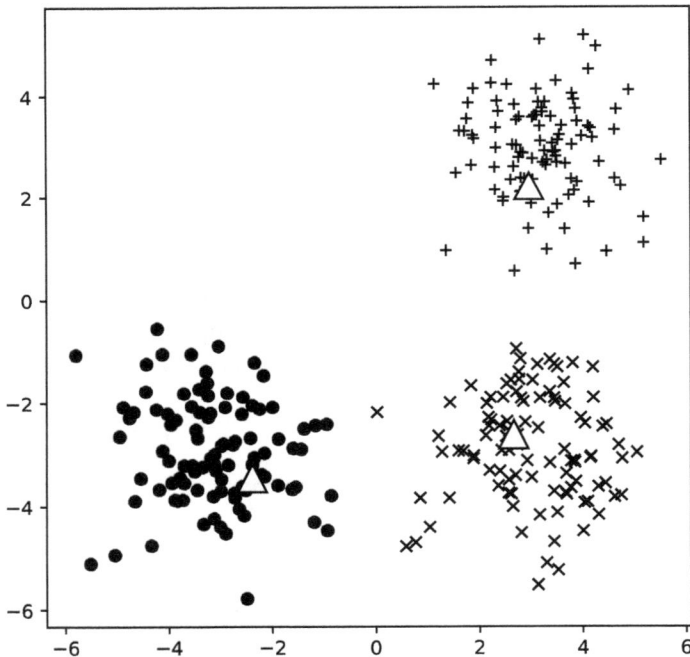

FIGURE 15.1: Clusters produced by the PP algorithm. The attractors are indicated by triangles.

The following code shows the application of the PP algorithm to the second synthetic dataset:

```
X, y = make_circles(n_samples=300, noise=0.05, factor=0.4,
    random_state=0)
ci, W = pp(X, s=10)
yhat = np.argmax(W, axis=1)
```

```
4  cm1 = createCM(y, yhat)
5  print(cm1)
```

The second synthetic dataset contains two circles. Each circle is considered as a cluster as the points in a circle are connected.

Executing the above block of code gives the following output:

```
1        38      68
2  0      0     150
3  1    150       0
```

From the output, we see that all records are clusterized correctly.

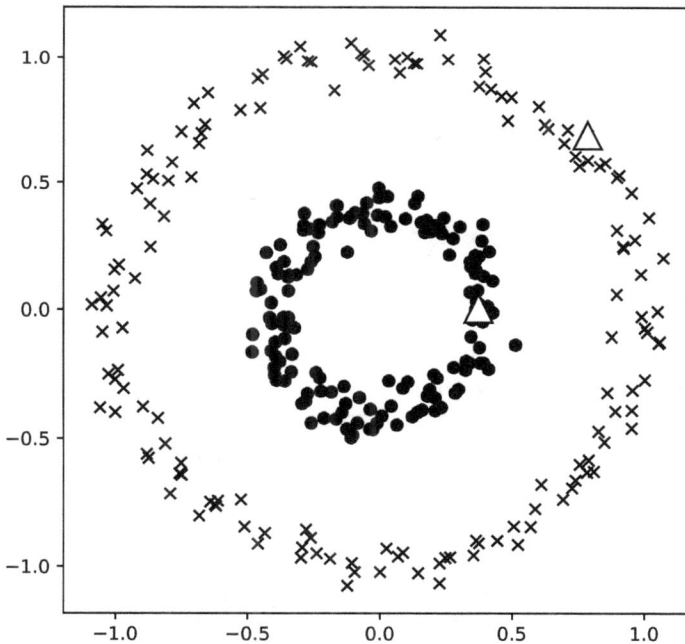

FIGURE 15.2: Clusters produced by the PP algorithm. The attractors are indicated by triangles.

Figure 15.2 shows the clusters obtained by applying the PP algorithm to the second synthetic dataset. From the plot, we see that the attractors are not located at the centers of the clusters.

Now let us apply the PP algorithm to the Iris dataset with default values of the parameters:

```
1 iris = fetch_ucirepo(id=53)
2 X = iris.data.features
3 y = iris.data.targets
4
5 ci, W = pp(X)
6 yhat = np.argmax(W, axis=1)
7 cm1 = createCM(y, yhat)
8 print(cm1)
```

Executing the above block of code gives the following output:

	0	37	86	94	120	149
Iris-setosa	29	21	0	0	0	0
Iris-versicolor	0	0	11	34	1	4
Iris-virginica	0	0	0	1	35	14

From the confusion matrix, we see that the six clusters are obtained when default values of the parameters are used.

To reduce the number of clusters, we can increase the parameter s. To do that, we can use the following code:

```
1 ci, W = pp(X, s=10)
2 yhat = np.argmax(W, axis=1)
3 cm1 = createCM(y, yhat)
4 print(cm1)
```

In the above code, we use 10 for the parameter s. Executing the above block of code gives the following output:

	17	99
Iris-setosa	50	0
Iris-versicolor	0	50
Iris-virginica	0	50

The output shows that two clusters are obtained.

To test the sparse implementation of the PP algorithm, we use the following code:

```
1 ci, W = pps(X)
2 yhat = np.argmax(W, axis=1)
3 cm1 = createCM(y, yhat)
4 print(cm1)
5 W
```

Executing the above block of code in Spyder gives the following output:

```
1                      0     37    86    94    120   149
2  Iris-setosa        29     21     0     0      0     0
3  Iris-versicolor     0      0    11    34      1     4
4  Iris-virginica      0      0     0     1     35    14
5  (150, 150)
6  Out[253]:
7  <Compressed Sparse Row sparse array of dtype 'float64'
8    with 9751 stored elements and shape (150, 150)>
```

The sparse implementation produced the same confusion matrix as the full implementation. The sparse stochastic matrix only saved 9751 elements. The full stochastic matrix contains $150 \times 150 = 22500$ elements. The sparse implementation saves lots of memory space.

15.4 Summary

In this chapter, we introduced and implemented the Probability Propagation (PP) algorithm [98], a graph-based clustering method capable of identifying arbitrarily shaped clusters. We provided two implementations: one utilizing a full stochastic matrix and another employing a sparse stochastic matrix. The sparse implementation is particularly useful for handling large datasets efficiently.

16

A Spectral Clustering Algorithm

Spectral clustering algorithms are graph-based clustering algorithms that are capable of identifying clusters with arbitrary shapes. In this chapter, we introduce and implement a spectral clustering algorithm.

16.1 Description of the Algorithm

Spectral clustering refers to techniques that utilize the eigenvalues (i.e., the spectrum) of a dataset's similarity matrix to group data. A typical spectral clustering algorithm consists of three steps [2, Chapter 8]: first, construct a similarity matrix of the dataset; second, use the eigenvalues to map the data into a feature space where clusters are more obvious; third, apply a classical clustering algorithm (e.g., k-means) to clusterize the transformed data.

There are multiple ways to construct a similarity matrix from a dataset. Here we introduce two methods: the ϵ-neighbor method and the Gaussian method. Let $D = \{\mathbf{x}_0, \mathbf{x}_1, \ldots, \mathbf{x}_{n-1}\}$ be a dataset containing n records. Under the ϵ-neighbor method, the similarity matrix S is constructed as follows:

$$s_{ij} = \begin{cases} 1, & \text{if } d(\mathbf{x}_i, \mathbf{x}_j) < \epsilon, \\ 0, & \text{if } d(\mathbf{x}_i, \mathbf{x}_j) \geq \epsilon, \end{cases} \tag{16.1}$$

where $\epsilon > 0$ is a parameter, $d(\cdot, \cdot)$ is a distance function, and s_{ij} is the (i, j)-th entry of S. Under the Gaussian method, the similarity matrix S is constructed as follows:

$$s_{ij} = \exp\left(-\frac{d(\mathbf{x}_i, \mathbf{x}_j)^2}{2\delta^2}\right), \tag{16.2}$$

where δ is a bandwidth parameter.

After constructing the similarity matrix S, we define the following diagonal matrix E:

$$e_{ij} = \begin{cases} \sum_{j=0}^{n-1} s_{ij}, & \text{if } i == j, \\ 0, & \text{if } i \neq j. \end{cases}$$

DOI: 10.1201/9781003592648-16

The diagonal matrix E is called a degree matrix. Then we construct the symmetrically normalized graph Laplacian:

$$L = E^{-1/2}(E - S)E^{-1/2} = I - E^{-1/2}SE^{-1/2}. \qquad (16.3)$$

After the graph Laplacian L is constructed, we obtain the eigenvectors corresponding to the h smallest eigenvalues of L, where h is a parameter. Let U be the matrix formed by the h eigenvectors. Each eigenvector is a column of U. Next, we apply the k-means algorithm to the rows of U to identify clusters.

16.2 Implementation

The spectral clustering algorithm described in the previous section can be implemented in Python with a few lines of code. Since the spectral clustering algorithm consists of three major steps, we implement each step as a function. We only need to implement the first two steps as the third step involves application of an existing clustering algorithm such as k-means.

Constructing the similarity matrix can be done in several ways. The following function implements the ϵ-neighbor method:

```
1  def epsilonneighbor(dm, epsilon=None):
2      n = dm.shape[0]
3      if epsilon is None:
4          epsilon = np.mean(dm)/5
5      S = np.zeros((n,n))
6      S[dm < epsilon] = 1
7      return S
```

The input to the above function is a distance matrix and a parameter, which is the ϵ. If the parameter is not provided, it will be estimated to be one fifth of the mean distance.

The following function implements the Gaussian method:

```
1  def gaussian(dm, delta=None):
2      if delta is None:
3          delta = np.mean(dm)/5
4      S = np.exp(-np.square(dm/delta)/2)
5      return S
```

The Gaussian method constructs a fully connected similarity matrix. This function also has two inputs: a distance matrix and a bandwidth parameter. If the bandwidth parameter is not provided, it will be estimated to be one fifth of the mean distance.

The following function implements the second step:

```
1  def spectral(S, h=3):
2      L = laplacian(S, normed=True)
3      ev = np.linalg.eigh(L)
4      ind = np.argsort(ev[0])
5      U = np.asanyarray(ev[1][:,ind[0:h]], float)
6      U = U/np.sqrt(1e-8+np.sum(np.square(U), axis=1,
           keepdims=True))
7      return U
```

This function takes a similarity matrix as an input and returns a matrix of selected eigenvectors. This function also has a parameter used to specify how many eigenvectors will be used. These eigenvectors correspond to the smallest eigenvalues. The function `laplacian` from the SciPy library is used to create the Laplacian matrix. The function `eigh` from the NumPy library is used to calculate the eigenvalues and eigenvectors. Each row of the resulting matrix U is normalized. To prevent division by zero, we add a small constant to the normalization operation.

16.3 Examples

In this section, we illustrate the spectral clustering algorithm with two synthetic datasets and the Iris dataset. The first synthetic dataset contains spherically shaped clusters. The second synthetic dataset contains chain-like clusters.

The examples in this section requires the following libraries and functions:

```
1  import numpy as np
2  import matplotlib.pyplot as plt
3  from scipy.spatial.distance import pdist, squareform
4  from scipy.sparse.csgraph import laplacian
5  from sklearn.datasets import make_blobs, make_circles
6  from sklearn.cluster import SpectralClustering
7  from ucimlrepo import fetch_ucirepo
8  from dcutil import createCM
9  from kmeans import kmeans2
```

We run the above block of code first before all other code presented in this section.

The following block of code shows how to apply the spectral clustering algorithm to the first synthetic dataset:

```
1  centers = [[3, 3], [-3, -3], [3, -3]]
2  X, y = make_blobs(n_samples=300, centers=centers,
       cluster_std=1, random_state=1)
```

```
3
4  dm = squareform(pdist(X))
5
6  S = epsilonneighbor(dm)
7  U = spectral(S, h=4)
8  bcm, bcc, bov, biters = kmeans2(U, k=3)
9  cm2 = createCM(y, bcm)
10 print([bov, biters])
11 print(cm2)
```

In the above code, we use the ϵ-neighbor method to construct the similarity matrix. We use 4 eigenvectors to create the new dataset. Executing the above block of code gave the following output:

```
1  [52.43913620381592, 5]
2  [51.989854771578365, 3]
3  [2.8784567908596466, 2]
4  [52.43913620381592, 5]
5  [2.8784567908596466, 3]
6  [2.8784567908596466, 3]
7  [2.8784567908596466, 3]
8  [52.43913620381592, 4]
9  [2.8784567908596466, 2]
10 [52.43913620381592, 4]
11 [2.8784567908596466, 2]
12        0    1    2
13 0    100    0    0
14 1      0    0  100
15 2      0  100    0
```

The output confirms that all clusters were correctly identified.

To visualize the dataset, the similarity matrix, the transformed data, and the clustering results, we execute the following code:

```
1  fig, ax = plt.subplots(2, 2, figsize=(8, 8))
2  ax[0,0].scatter(X[:,0], X[:,1], color="black")
3  ax[0,1].scatter(X[:,0], X[:,1], color="black")
4  ind = np.where(S > 0)
5  for i in range(ind[0].shape[0]):
6      ij = [ind[0][i], ind[1][i]]
7      ax[0,1].plot(X[ij,0], X[ij,1], '-', color="grey")
8  ax[1,0].scatter(U[:,0], U[:,1], color="black")
9  for i in range(3):
10     center = bcc[i,:]
11     ax[1,0].plot(center[0], center[1], "^", markerfacecolor
           ="white",
12         markeredgecolor="black", markersize=15)
13 markers = ["x", "o", "+"]
14 for i in range(3):
```

```
15    members = bcm == i
16    ax[1,1].plot(X[members, 0], X[members, 1], markers[i],
         color="black")
17 ax[0,0].set_title("D")
18 ax[0,1].set_title("S")
19 ax[1,0].set_title("U")
20 ax[1,1].set_title("CM")
```

Figure 16.1 shows the resulting plot. In the top right subplot, we see the connections of the points that are relatively close to each other. The bottom left subplot shows the first two eigenvectors as we cannot visualize all the four eigenvectors used by the k-means algorithm.

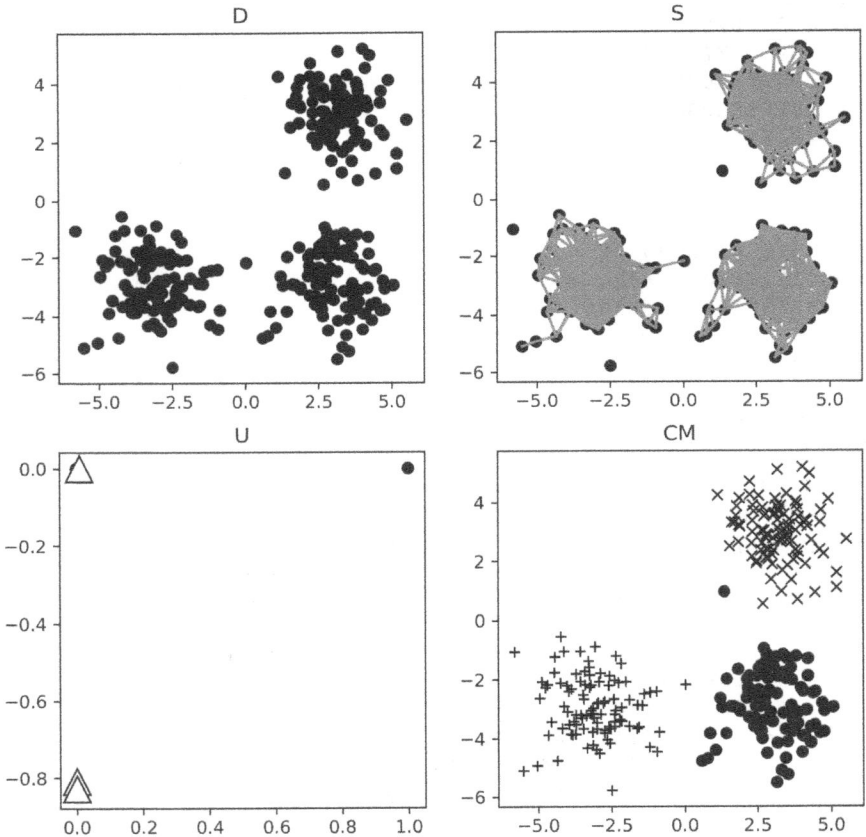

FIGURE 16.1: The first synthetic dataset and results obtained by the spectral clustering algorithm. Subplots with titles D, S, U, and CM correspond to the original data, the similarity matrix, the first two columns of U, and the clustering results.

To see the result of the similarity matrix constructed by the Gaussian method, we run the following block of code:

```
S = gaussian(dm)
U = spectral(S, h=4)
bcm, bcc, bov, biters = kmeans2(U, k=3)
cm2 = createCM(y, bcm)
print([bov, biters])
print(cm2)
```

Executing the above block of codes produced the following output:

```
[46.503105388712655, 2]
[46.503105388712655, 6]
[46.503105388712655, 4]
[46.503105388712655, 8]
[46.503105388712655, 8]
[46.503105388712655, 4]
[46.503105388712655, 8]
[46.503105388712655, 8]
[46.503105388712655, 8]
[46.503105388712655, 7]
[46.503105388712655, 2]
        0    1    2
0    100    0    0
1      0    1   99
2      1   99    0
```

From the output, we see that two records were misclassified.

The spectral clustering algorithm is also implemented in the scikit-learn library. To see the results of applying the scikit-learn algorithm to the first synthetic dataset, we run the following code:

```
clustering = SpectralClustering(n_clusters=3,
        assign_labels='discretize',
        random_state=0).fit(X)
cm = createCM(y, clustering.labels_)
print(cm)
```

After executing the above block of code, we see the following output:

```
        0    1    2
0    100    0    0
1      0    0  100
2      1   99    0
```

The output shows that only one record was misclassified.

Now we test the spectral clustering algorithm on the second synthetic dataset, which contains two circles. To do that, we use the following code:

```
1  X, y = make_circles(n_samples=300, noise=0.05, factor=0.4,
       random_state=0)
2
3  dm = squareform(pdist(X))
4
5  S = epsilonneighbor(dm)
6  U = spectral(S, h=4)
7  bcm, bcc, bov, biters = kmeans2(U, k=2)
8  cm2 = createCM(y, bcm)
9  print([bov, biters])
10 print(cm2)
```

Executing the above block of code produced the following output:

```
1  [99.62947526074446, 4]
2  [99.62947526074446, 6]
3  [99.62947526074446, 7]
4  [99.62947526074446, 4]
5  [99.62947526074446, 4]
6  [99.62947526074446, 3]
7  [99.62947526074446, 5]
8  [99.62947526074446, 4]
9  [99.62947526074446, 4]
10 [99.62947526074446, 6]
11 [99.62947526074446, 4]
12        0      1
13 0    150      0
14 1      0    150
```

The output shows that all records were classified correctly.

To visualize the results, we run the following block of code:

```
1  fig, ax = plt.subplots(2, 2, figsize=(8, 8))
2  ax[0,0].scatter(X[:,0], X[:,1], color="black")
3  ax[0,1].scatter(X[:,0], X[:,1], color="black")
4  ind = np.where(S > 0)
5  for i in range(ind[0].shape[0]):
6      ij = [ind[0][i], ind[1][i]]
7      ax[0,1].plot(X[ij,0], X[ij,1], '-', color="grey")
8  ax[1,0].scatter(U[:,0], U[:,1], color="black")
9  for i in range(2):
10     center = bcc[i,:]
11     ax[1,0].plot(center[0], center[1], "^", markerfacecolor
           ="white",
12         markeredgecolor="black", markersize=15)
13 markers = ["x", "o", "+"]
14 for i in range(2):
15     members = bcm == i
```

```
16        ax[1,1].plot(X[members, 0], X[members, 1], markers[i],
             color="black")
17  ax[0,0].set_title("D")
18  ax[0,1].set_title("S")
19  ax[1,0].set_title("U")
20  ax[1,1].set_title("CM")
```

The resulting plot produced by the above code is shown in Figure 16.2. The subplots show the original data, the connections in the similarity matrix, the first two eigenvectors, and the clustering results. The results show that the spectral clustering algorithm can find clusters of chain-like shapes.

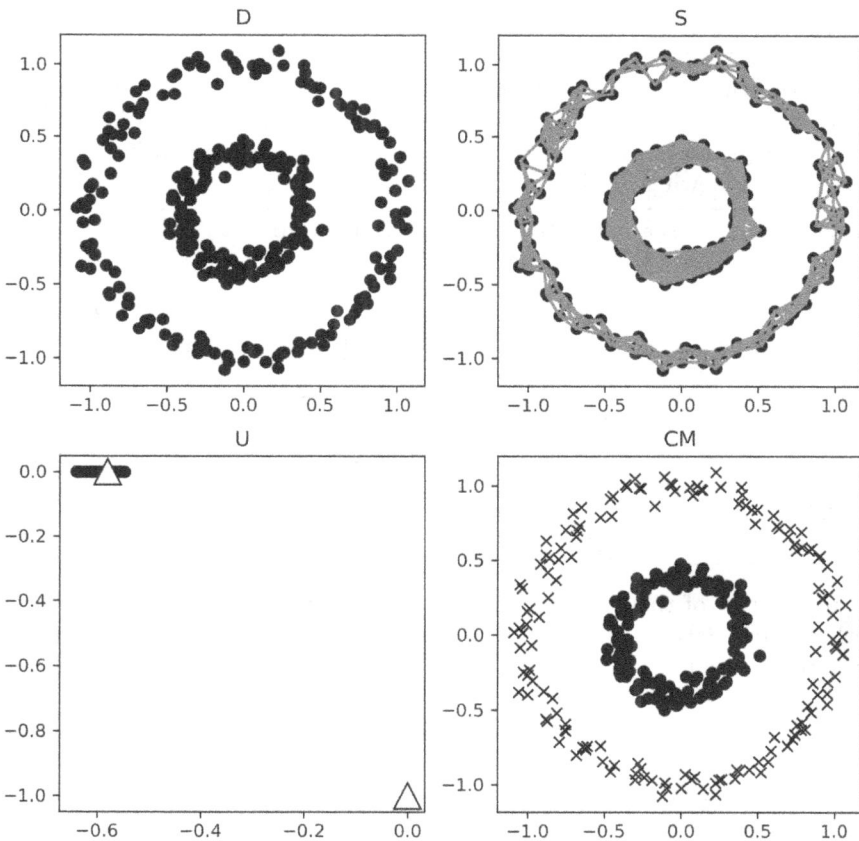

FIGURE 16.2: The second synthetic dataset and results obtained by the spectral clustering algorithm. Subplots with titles D, S, U, and CM correspond to the original data, the similarity matrix, the first two columns of U, and the clustering results.

We can also apply the scikit-learn function to perform spectral clustering on the second synthetic data. To do that, we run the following code:

```
clustering = SpectralClustering(n_clusters=2,
         assign_labels='discretize',
         random_state=0).fit(X)
cm = createCM(y, clustering.labels_)
print(cm)
```

Again, we use default values for many other parameters of the function. After executing the above block of code, we see the following output:

```
     0   1
0   74  76
1   70  80
```

The output indicates that with default parameters, the function fails to correctly separate the two clusters. To improve the results, we can increase the parameter **gamma** used by the function to calculate the similarity matrix. If we run the following block of code:

```
clustering = SpectralClustering(n_clusters=2,
         assign_labels='discretize',
         gamma = 20,
         random_state=0).fit(X)
cm = createCM(y, clustering.labels_)
print(cm)
```

we will get the following output:

```
     0    1
0   150    0
1    0   150
```

In the rest of this section, we illustrate the spectral clustering algorithm with the Iris dataset. To apply the spectral clustering algorithm to the Iris dataset, we use the following code:

```
X = iris.data.features
y = iris.data.targets

dm = squareform(pdist(X))

S = gaussian(dm)
U = spectral(S, h=3)
bcm, bcc, bov, biters = kmeans2(U, k=3, numrun=100)
cm2 = createCM(y, bcm)
print([bov, biters])
print(cm2)
```

In the above code, we use the Gaussian method to create the similarity matrix. Executing the above block of code gave the following output:

```
[46.67530552378307 , 6]
[12.181825121513388 , 8]
[12.181825121513388 , 4]
[12.19559744813257 , 5]
[12.181825121513388 , 8]
[12.19559744813257 , 4]
[12.181825121513388 , 5]
[12.181825121513388 , 8]
[12.181825121513388 , 10]
[12.181825121513388 , 11]
[12.181825121513388 , 8]
                      0   1   2
Iris-setosa           0  50   0
Iris-versicolor      48   0   2
Iris-virginica       12   0  38
```

From the output, we see that 14 records were misclassified.

To see the results of applying the scikit-learn algorithm to the Iris data, we use the following code:

```
clustering = SpectralClustering(n_clusters=3,
        assign_labels='discretize',
        random_state=0).fit(X)
cm = createCM(y, clustering.labels_)
print(cm)
```

We use default values for other parameters. Executing the above block of code gave the following output:

```
                 0   1   2
Iris-setosa      0  50   0
Iris-versicolor  2   0  48
Iris-virginica  37   0  13
```

The output shows that 15 records were misclassified.

16.4   Summary

In this chapter, we implemented a spectral clustering algorithm and demonstrated its effectiveness using two synthetic datasets and the Iris dataset. Spectral clustering is closely related to kernel-based clustering. Notably, the kernel k-means algorithm is mathematically equivalent to spectral clustering [60]. More information about spectral clustering can be found in [196], [246], and [2, Chapter 8].

17

A Mean-Shift Algorithm

Mean-shift algorithms are clustering methods used to identify regions with high data density. There exist multiple variations of the mean-shift algorithm. In this chapter, we introduce a simple mean-shift algorithm and its implementation.

17.1 Description of the Algorithm

Mean-shift algorithms have a long history. According to [126, Chapter 18], the mean-shift algorithm and the term "mean shift" were perhaps first proposed by Fukunaga and Hostetler [85] in 1975.

Mean-shift clustering algorithms can be derived from a kernel density estimate. Let $D = \{\mathbf{x}_0, \mathbf{x}_1, \ldots, \mathbf{x}_{n-1}\}$ be a set of n multivariate data points. Then a kernel density estimate can be defined as

$$p(\mathbf{x}) = \frac{1}{n} \sum_{i=0}^{n-1} K\left(\frac{\|\mathbf{x} - \mathbf{x}_i\|}{\sigma}\right), \qquad (17.1)$$

where $K(\cdot)$ is a kernel function, $\sigma > 0$ is a bandwidth parameter, and $\|\cdot\|$ is the L^2 norm or Euclidean distance.

To identify a mode in the kernel density estimate, we can take the derivative of $p(\mathbf{x})$ with respect to $\mathbf{x}$ and set it to zero. By doing so, we get the following equation:

$$p'(\mathbf{x}) = \frac{1}{n} \sum_{i=0}^{n-1} K'\left(\frac{\|\mathbf{x} - \mathbf{x}_i\|}{\sigma}\right) \frac{(\mathbf{x} - \mathbf{x}_i)}{2\sigma\|\mathbf{x} - \mathbf{x}_i\|} = 0, \qquad (17.2)$$

solving which gives

$$\mathbf{x} = \sum_{i=0}^{n-1} \frac{K'\left(\dfrac{\|\mathbf{x} - \mathbf{x}_i\|}{\sigma}\right) \dfrac{1}{\|\mathbf{x} - \mathbf{x}_i\|}}{\sum_{j=0}^{n-1} K'\left(\dfrac{\|\mathbf{x} - \mathbf{x}_j\|}{\sigma}\right) \dfrac{1}{\|\mathbf{x} - \mathbf{x}_j\|}} \mathbf{x}_i. \qquad (17.3)$$

DOI: 10.1201/9781003592648-17

Equation (17.3) can be used to formulate the following iterative update of the mode estimate:

$$\mathbf{x}^{(t+1)} = \sum_{i=0}^{n-1} \frac{K'\left(\dfrac{\|\mathbf{x}^{(t)} - \mathbf{x}_i\|}{\sigma}\right) \dfrac{1}{\|\mathbf{x}^{(t)} - \mathbf{x}_j\|}}{\sum_{j=0}^{n-1} K'\left(\dfrac{\|\mathbf{x}^{(t)} - \mathbf{x}_j\|}{\sigma}\right) \dfrac{1}{\|\mathbf{x}^{(t)} - \mathbf{x}_j\|}} \mathbf{x}_i, \quad t = 0, 1, \dots, \quad (17.4)$$

where $\mathbf{x}^{(0)}$ is some initial guess of the mode.

Different kernel functions lead to different mean-shift algorithms. Common kernel functions include the Gaussian kernel function and the Epanechnikov kernel function. The Gaussian kernel function is defined by

$$K(u) = \frac{1}{\sqrt{2\pi}} \exp\left(-\frac{u^2}{2}\right).$$

The derivative of the Gaussian kernel function is

$$K'(u) = -\frac{1}{\sqrt{2\pi}} \exp\left(-\frac{u^2}{2}\right) u$$

The Epanechnikov kernel function is defined as

$$K(u) = \begin{cases} \dfrac{3}{4}(1 - u^2), & \text{if } |u| \leq 1, \\ 0, & \text{if } |u| > 1. \end{cases}$$

The Epanechnikov kernel is also called the parabolic kernel. The derivative of the Epanechnikov kernel function is

$$K'(u) = \begin{cases} -\dfrac{3}{2}u, & \text{if } |u| < 1 \\ 0, & \text{if otherwise.} \end{cases}$$

When using the Gaussian kernel, the mean-shift iteration is expressed as

$$\mathbf{x}^{(t+1)} = \sum_{i=0}^{n-1} \frac{\exp\left(-\dfrac{\|\mathbf{x}^{(t)} - \mathbf{x}_i\|^2}{2\sigma^2}\right)}{\sum_{j=0}^{n-1} \exp\left(-\dfrac{\|\mathbf{x}^{(t)} - \mathbf{x}_j\|^2}{2\sigma^2}\right)} \mathbf{x}_i, \quad t = 0, 1, \dots, \quad (17.5)$$

Where using the parabolic kernel, the mean-shift iteration becomes

$$\mathbf{x}^{(t+1)} = \sum_{i=0}^{n-1} \frac{I_{\{\|\mathbf{x}^{(t)} - \mathbf{x}_i\| < \sigma\}}}{\sum_{j=0}^{n-1} I_{\{\|\mathbf{x}^{(t)} - \mathbf{x}_j\| < \sigma\}}} \mathbf{x}_i, \quad t = 0, 1, \dots, \quad (17.6)$$

where I is an indicator function. The above iteration can be expressed as

$$\mathbf{x}^{(t+1)} = \frac{1}{|N(\mathbf{x}^{(t)})|} \sum_{\mathbf{y} \in N(\mathbf{x}^{(t)})} \mathbf{y},$$

where $N(\mathbf{x}^{(t)})$ is the set of data points that have less than δ distance from $\mathbf{x}^{(t)}$, i.e.,

$$N(\mathbf{x}^{(t)}) = \{\mathbf{y} \in D : \|\mathbf{y} - \mathbf{x}^{(t)}\| < \delta\}.$$

Mean-shift iterations can be represented in matrix form. Suppose that $\mathbf{x}_i$'s are column vectors. Let

$$X = \begin{pmatrix} \mathbf{x}_0^T \\ \mathbf{x}_1^T \\ \vdots \\ \mathbf{x}_{n-1}^T \end{pmatrix}$$

be the $n \times d$ data matrix, where d is the dimensionality of the data. Let $Z = X$ be the matrix containing the initial guesses of the modes. For the Gaussian kernel, we let W be a matrix defined by

$$w_{ij} = \exp\left(-\frac{\|\mathbf{z}_i - \mathbf{x}_j\|^2}{2\sigma^2}\right), \quad i,j = 0,1,\ldots,n-1.$$

Let E be the diagonal matrix defined by

$$e_{ij} = \begin{cases} \sum_{j=0}^{n-1} w_{ij}, & \text{if } i = j \\ 0, & \text{if } i \neq j, \end{cases} \quad i,j = 0,1,\ldots,n-1.$$

Then the mode matrix Z can be updated as follows:

$$Z \leftarrow WE^{-1}X. \tag{17.7}$$

Equation (17.7) can also be used for the parabolic kernel. When the parabolic kernel is used, the weight matrix W is calculated as

$$w_{ij} = \begin{cases} 1, & \text{if } \|\mathbf{z}_i - \mathbf{x}_j\| < \sigma, \\ 0, & \text{if } \|\mathbf{z}_i - \mathbf{x}_j\| \geq \sigma, \end{cases} \quad i,j = 0,1,\ldots,n-1.$$

Mean-shift algorithms require one parameter: the bandwidth. The number of clusters is not required. The bandwidth determines how many clusters the algorithm will produce. A larger bandwidth leads to fewer clusters.

The iterative process of a mean-shift algorithm terminates when either a maximum number of iterations is reached or mode changes fall within a specified tolerance. Some modes found by a mean-shift algorithm can be close to each other. As a result, the modes need to be postprocessed to merge modes that are close to each other. After a final set of modes is determined, the data points are assigned to their closest modes.

17.2 Implementation

In this section, we implement two mean-shift algorithms. The first one uses the parabolic kernel function and the second one uses the Gaussian kernel function.

The first mean-shift algorithm is implemented as follows:

```python
def msparabolic(X, sigma=1, tol=1e-8, maxit=300):
    X = np.ascontiguousarray(X)
    Z = X.copy()
    iters = 1
    while iters < maxit:
        W = (pairwise_distances(Z, X) < sigma).astype(float
            )
        E = np.sum(W, axis=1, keepdims=True)
        Z_ = Z
        Z = (W/E) @ X
        iters += 1
        if np.max(np.abs(Z-Z_)) < tol:
            break
    n, d = X.shape
    clusterCenter = np.zeros((0,d))
    sind = np.argsort(E[:,0])[::-1]
    unassigned = [index.item() for index in sind if index >
        0]
    while len(unassigned) > 0:
        i = unassigned[0]
        dist = pairwise_distances(Z[i,:].reshape(1, -1), Z[
            unassigned,:])
        ind = np.where(dist[0,:] < sigma)[0]
        clusterCenter = np.append(clusterCenter, Z[i,:].
            reshape(1,-1), axis=0)
        for j in ind:
            sind[sind==unassigned[j]] = -1
        unassigned = [index.item() for index in sind if
            index > 0]
    dist = pairwise_distances(X, clusterCenter)
    clusterMembership = np.argmin(dist, axis=1)
    return clusterMembership, clusterCenter, Z, iters
```

In the above code, we use the matrix form to speed up the calculation. Lines 5–12 implement the mean-shift iteration given in Equation (17.7). Lines 13–26 are the postprocessing step. We sort the modes by density. The density of a mode is defined to be the number of data points near the mode. Modes that are within the bandwidth are grouped together. The function pairwise_distances from the scikit-learn library is used to calculate the pairwise distances between two datasets.

The mean-shift algorithm with the Gaussian kernel is implemented similarly as follows:

```python
def msgaussian(X, sigma=1, tol=1e-8, maxit=300):
    X = np.ascontiguousarray(X)
    Z = X.copy()
    iters = 1
```

```
 5    while iters < maxit:
 6        W = np.exp(-0.5*np.square(pairwise_distances(Z, X)/
              sigma))
 7        E = np.sum(W, axis=1, keepdims=True)
 8        Z_ = Z
 9        Z = (W/E) @ X
10        iters += 1
11        if np.max(np.abs(Z-Z_)) < tol:
12            break
13    n, d = X.shape
14    clusterCenter = np.zeros((0,d))
15    sind = np.argsort(E[:,0])[::-1]
16    unassigned = [index.item() for index in sind if index >
          0]
17    while len(unassigned) > 0:
18        i = unassigned[0]
19        dist = pairwise_distances(Z[i,:].reshape(1, -1), Z[
              unassigned,:])
20        ind = np.where(dist[0,:] < sigma)[0]
21        clusterCenter = np.append(clusterCenter, Z[i,:].
              reshape(1,-1), axis=0)
22        for j in ind:
23            sind[sind==unassigned[j]] = -1
24        unassigned = [index.item() for index in sind if
              index > 0]
25    dist = pairwise_distances(X, clusterCenter)
26    clusterMembership = np.argmin(dist, axis=1)
27    return clusterMembership, clusterCenter, Z, iters
```

The code is almost the same as that of the mean-shift algorithm with the parabolic kernel. The only difference is the calculation of the matrix W.

17.3 Examples

In this section, we illustrate the mean-shift algorithms with two synthetic datasets and the Iris dataset. The examples in this section require the following Python libraries and functions:

```
1  import numpy as np
2  import matplotlib.pyplot as plt
3  from scipy.spatial.distance import pdist
4  from sklearn.datasets import make_blobs, make_circles
5  from sklearn.metrics import pairwise_distances
6  from sklearn.cluster import MeanShift
7  from ucimlrepo import fetch_ucirepo
```

```
8 from dcutil import createCM
```

The following code shows the application of the mean-shift algorithm with the parabolic kernel to the first synthetic dataset:

```
1 centers = [[3, 3], [-3, -3], [3, -3]]
2 X, y = make_blobs(n_samples=300, centers=centers,
      cluster_std=1, random_state=1)
3
4 dm = pdist(X)
5 sigma = np.percentile(dm, 25)
6
7 yhat, cc, Z, iters = msparabolic(X, sigma=sigma)
8 cm1 = createCM(y, yhat)
9 print(cm1)
```

The bandwidth parameter used above is the 25th percentile of the distances among the different data points. Executing the above block of code gave the following output:

```
1       0    1    2
2 0   100    0    0
3 1     0    1   99
4 2     1   99    0
```

From the confusion matrix, we see that two data points were misclassified. We can plot the clustering results by using the following code:

```
1 fig, ax = plt.subplots(1, 1, figsize=(6, 6))
2 ax.scatter(X[:,0], X[:,1], color="white")
3 for i in range(cc.shape[0]):
4     ax.plot(cc[i,0], cc[i,1], "^", markerfacecolor="white",
5             markeredgecolor="black", markersize=20,
              markeredgewidth=2)
6 for i in range(X.shape[0]):
7     ax.text(X[i,0], X[i, 1], str(yhat[i]), fontsize=12)
```

The above code plots the cluster centers (i.e., modes) by triangles and the data points by their indices of clusters. Figure 17.1 shows the clustering results plotted by the above code.

The mean-shift algorithm with the parabolic kernel is also provided by the scikit-learn library. To apply this algorithm to the first synthetic dataset, we use the following code:

```
1 clustering = MeanShift(bandwidth=sigma).fit(X)
2 cm = createCM(y, clustering.labels_)
3 print(cm)
```

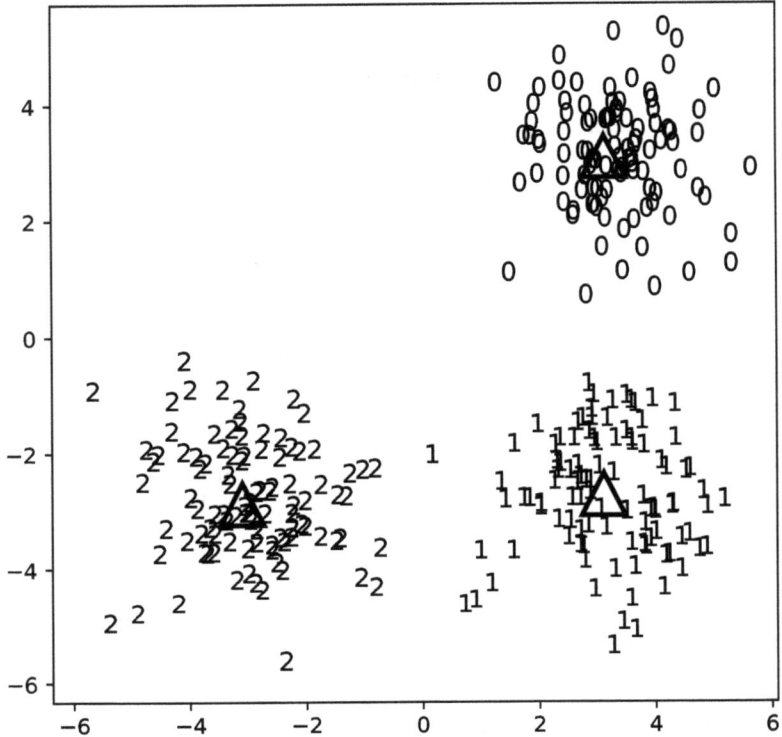

FIGURE 17.1: Clusters produced by the mean-shift algorithm with the parabolic kernel. Modes are indicated by triangles.

The bandwidth parameter is set to the same value used previously. Executing the above block of code gave the following output:

```
        0    1    2
0     100    0    0
1       0    1   99
2       1   99    0
```

The scikit-learn algorithm and our implementation produced the same results.

To test the mean-shift algorithm with the Gaussian kernel on the first synthetic data, we use the following code:

```
yhat, cc, Z, iters = msgaussian(X, sigma=sigma)
cm1 = createCM(y, yhat)
print(cm1)
```

Executing the above block of code gave the following output:

```
        0    1    2
0  0     0   100    0
1  1     1     0   99
2  2    99     1    0
```

From the output, we see that the mean-shift algorithm with the Gaussian kernel produced the same results as the one with the parabolic kernel.

The second synthetic dataset contains two circular clusters. The following code shows the application of the mean-shift algorithm with the parabolic kernel to the second synthetic dataset:

```
X, y = make_circles(n_samples=300, noise=0.05, factor=0.4,
    random_state=0)
X = np.round(X, decimals=4)

dm = pdist(X)
sigma = np.percentile(dm, 15)

yhat, cc, Z, iters = msparabolic(X, sigma=sigma)
cm1 = createCM(y, yhat)
print(cm1)
```

This time we use the 15th percentile of the distances as the bandwidth. If we use the 50th percentile of the distances as the bandwidth, we will get just one cluster. To produce more clusters, we need to reduce the bandwidth parameter. Executing the above block of code gave the following output:

```
        0    1    2    3
0   38   33   47   32
1  149    0    0    1
```

The resulting confusion matrix shows that four clusters were produced. The clustering results are also shown in Figure 17.2. From the figure, we see that the inner circle and part of the outer circle formed a a cluster. The other part of the outer circle was divided into three clusters.

To apply the scikit-learn mean-shift algorithm to the second synthetic dataset, we use the following code:

```
clustering = MeanShift(bandwidth=sigma).fit(X)
cm = createCM(y, clustering.labels_)
print(cm)
```

Executing the above block of code gave the following output:

```
        0    1    2    3
0   38   33   47   32
1  149    0    0    1
```

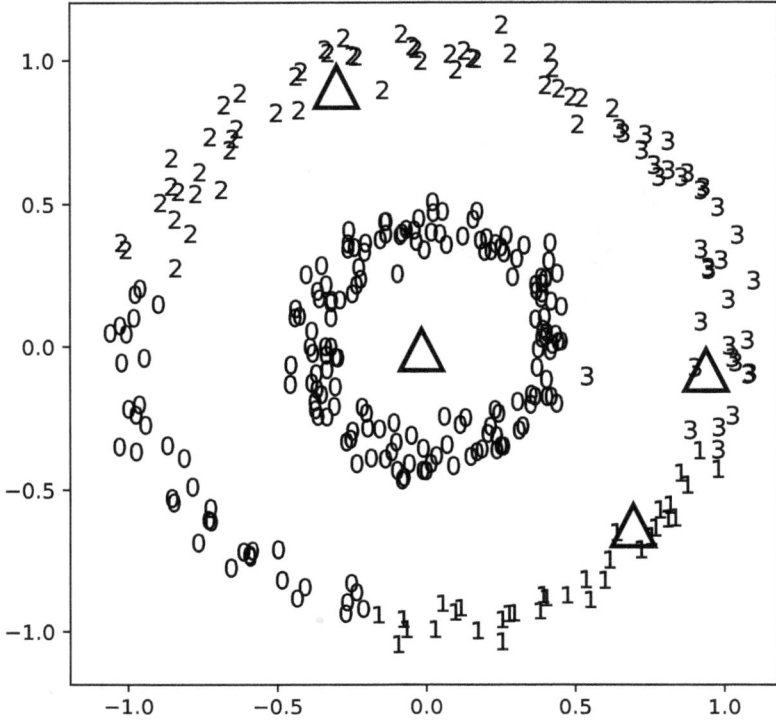

FIGURE 17.2: Clusters produced by the mean-shift algorithm with the parabolic kernel. Modes are indicated by triangles.

We see that our implementation and the scikit-learn implementation produced the same results.

To apply the mean-shift algorithm with the Gaussian kernel to the second synthetic dataset, we use the following code:

```
yhat, cc, Z, iters = msgaussian(X, sigma=sigma)
cm1 = createCM(y, yhat)
print(cm1)
yhat, cc, Z, iters = msgaussian(X, sigma=sigma/2)
cm1 = createCM(y, yhat)
print(cm1)
```

We run the mean-shift algorithm with different values for the bandwidth parameter. Executing the above block of code gave the following output:

```
        0
0    150
1    150
```

	0	1
0	76	74
1	76	74

When the bandwidth was set to the 15th percentile of the distances, one cluster was produced. When the bandwidth parameter was reduced to be half of the 15th percentile of the distances, two clusters were produced. The two clusters are shown in Figure 17.3. From the figure, we see that the two circles were split into two clusters.

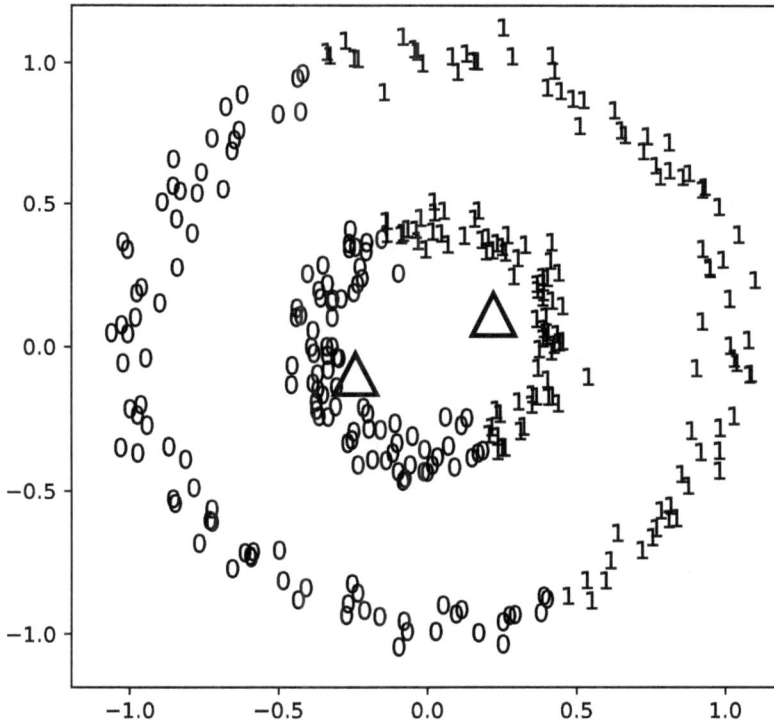

FIGURE 17.3: Clusters produced by the mean-shift algorithm with the Gaussian kernel. Modes are indicated by triangles.

Now let us apply the mean-shift algorithms to the Iris dataset. To apply the mean-shift algorithm with the parabolic kernel to the Iris dataset, we use the following code:

```
iris = fetch_ucirepo(id=53)
X = iris.data.features
y = iris.data.targets
```

```
5 dist = pdist(X)
6 sigma = np.percentile(dist, 15)
7 yhat, cc, Z, iters = msparabolic(X, sigma=sigma)
8 cm1 = createCM(y, yhat)
9 print(cm1)
```

In the above code, we first load the Iris data from the UCI machine learning repository. Then we set the bandwidth parameter to be the 15th percentile of the distances. Executing the above block of code gave the following output:

```
1                    0   1   2   3   4
2 Iris-setosa       50   0   0   0   0
3 Iris-versicolor    0  47   3   0   0
4 Iris-virginica     0   5  31  12   2
```

The output shows that five clusters were produced.

The following code shows the application of the scikit-learn mean-shift algorithm to the Iris dataset:

```
1 clustering = MeanShift(bandwidth=sigma).fit(X)
2 cm = createCM(y, clustering.labels_)
3 print(cm)
```

Executing the above block of code produced the following output:

```
1                    0   1   2   3   4
2 Iris-setosa       50   0   0   0   0
3 Iris-versicolor    0  47   3   0   0
4 Iris-virginica     0   5  31  12   2
```

From the output, we see that our implementation and the scikit-learn algorithm produced the same output.

To test the mean-shift algorithm with the Gaussian kernel on the Iris data, we run the following code:

```
1 yhat, cc, Z, iters = msgaussian(X, sigma=sigma)
2 cm1 = createCM(y, yhat)
3 print(cm1)
4 yhat, cc, Z, iters = msgaussian(X, sigma=sigma/3)
5 cm1 = createCM(y, yhat)
6 print(cm1)
```

Executing the above block of code gave the following output:

```
1                    0   1
2 Iris-setosa        0  50
3 Iris-versicolor   49   1
4 Iris-virginica    50   0
5                    0   1   2   3   4   5   6   7   8   9
```

6	Iris-setosa	50	0	0	0	0	0	0	0	0	0
7	Iris-versicolor	0	22	19	4	0	0	5	0	0	0
8	Iris-virginica	0	0	0	14	23	6	0	4	2	1

When the bandwidth was set to the 15th percentile of the distances, two clusters were produced. When the bandwidth was reduced to one third of the previous value, ten clusters were produced. Several clusters contained only a few data points.

Our experiments indicate that mean-shift algorithms perform well on datasets with spherical clusters. For datasets with chain-like structures, mean-shift algorithms may fragment these clusters into multiple smaller groups. Our results further highlight that mean-shift algorithms – particularly those using the Gaussian kernel – are highly sensitive to the bandwidth parameter.

17.4 Summary

In this chapter, we introduced and implemented the mean-shift algorithms. The mean-shift algorithm with the parabolic kernel is also implemented in the scikit-learn library. Our implementation is based on that of the scikit-learn library. However, the scikit-learn version is significantly more sophisticated than the one presented in this chapter. For more information on mean-shift algorithms, readers may refer to [36], [45], and [126, Chapter 18].

Bibliography

[1] Andreas Adolfsson, Margareta Ackerman, and Naomi C. Brownstein. To cluster, or not to cluster: An analysis of clusterability methods. *Pattern Recognition*, 88:13–26, April 2019.

[2] Charu C. Aggarwal and Chandan K. Reddy, editors. *Data Clustering: Algorithms and Applications*. Chapman and Hall/CRC, Boca Raton, FL, September 2014.

[3] Charu C. Aggarwal and Haixun Wang. A survey of clustering algorithms for graph data. In *Managing and Mining Graph Data*, pages 275–301. Springer US, 2010.

[4] Charu C. Aggarwal, Joel L. Wolf, Philip S. Yu, Cecilia Procopiuc, and Jong Soo Park. Fast algorithms for projected clustering. *ACM SIGMOD Record*, 28(2):61–72, June 1999.

[5] Charu C. Aggarwal and Philip S. Yu. Finding generalized projected clusters in high dimensional spaces. *ACM SIGMOD Record*, 29(2):70–81, May 2000.

[6] Saeed Aghabozorgi, Ali Seyed Shirkhorshidi, and Teh Ying Wah. Time-series clustering – a decade review. *Information Systems*, 53:16–38, October 2015.

[7] R. Agrawal, J. Gehrke, D. Gunopulos, and P. Raghavan. Automatic subspace clustering of high dimensional data for data mining applications. In *SIGMOD Record ACM Special Interest Group on Management of Data*, pages 94–105, New York, NY, USA, 1998. ACM Press.

[8] Amir Ahmad and Shehroz S. Khan. Survey of state-of-the-art mixed data clustering algorithms. *IEEE Access*, 7:31883–31902, 2019.

[9] Khaled S. Al-Sultan and Chawki A. Fedjki. A tabu search-based algorithm for the fuzzy clustering problem. *Pattern Recognition*, 30(12):2023–2030, December 1997.

[10] M.R. Anderberg. *Cluster Analysis for Applications*. Academic Press, New York, NY, 1973.

[11] William Steinbrunn Andras Janosi. Heart disease, 1989.

[12] David Báez-López and David Alfredo Báez Villegas. *Introduction to Python With Applications in Optimization, Image and Video Processing, and Machine Learning.* CRC Press, Boca Raton, FL, 2024.

[13] C.F. Banfield. Algorithm AS 102: Ultrametric distances for a single linkage dendrogram. *Applied Statistics*, 25(3):313, 1976.

[14] Jeffrey D. Banfield and Adrian E. Raftery. Model-based gaussian and non-gaussian clustering. *Biometrics*, 49(3):803, September 1993.

[15] A. Baraldi and P. Blonda. A survey of fuzzy clustering algorithms for pattern recognition. i. *IEEE Transactions on Systems, Man and Cybernetics, Part B (Cybernetics)*, 29(6):778–785, 1999.

[16] A. Baraldi and P. Blonda. A survey of fuzzy clustering algorithms for pattern recognition. ii. *IEEE Transactions on Systems, Man and Cybernetics, Part B (Cybernetics)*, 29(6):786–801, 1999.

[17] J. Basak and R. Krishnapuram. Interpretable hierarchical clustering by constructing an unsupervised decision tree. *IEEE Transactions on Knowledge and Data Engineering*, 17(1):121–132, January 2005.

[18] R. Bellman, R. Kalaba, and L.A. Zadeh. Abstraction and pattern classification. *Journal of Mathematical Analysis and Applications*, 2:581–586, 1966.

[19] Richard Bellman. A note on cluster analysis and dynamic programming. *Mathematical Biosciences*, 18(3-4):311–312, dec 1973.

[20] K.S. Beyer, J. Goldstein, R. Ramakrishnan, and U. Shaft. When is "nearest neighbor" meaningful? In C. Beeri and P. Buneman, editors, *Proceedings of 7th International Conference on Database Theory*, volume 1540 of *Lecture Notes in Computer Science*, pages 217–235. Springer, 1999.

[21] J.C. Bezdek. *Fuzzy mathematics in pattern classification.* PhD thesis, Cornell University, Ithaca, NY, April 1974.

[22] J.C. Bezdek. *Pattern Recognition with Fuzzy Objective Function Algorithms.* Plenum, New York, 1981.

[23] Panthadeep Bhattacharjee and Pinaki Mitra. A survey of density based clustering algorithms. *Frontiers of Computer Science*, 15(1), 2020.

[24] D.A. Binder. Bayesian cluster analysis. *Biometrika*, 65(1):31–38, April 1978.

[25] L. Bobrowski and J.C. Bezdek. c-means clustering with the l_1 and l_∞ norms. *IEEE Transactions on Systems, Man, and Cybernetics*, 21(3):545–554, 1991.

[26] H.H. Bock. *Probabilistic Aspects in Cluster Analysis*, pages 12–44. Springer Berlin Heidelberg, 1989.

[27] Hans H. Bock. Probabilistic models in cluster analysis. *Computational Statistics & Data Analysis*, 23(1):5–28, November 1996.

[28] Hans-Hermann Bock. Clustering methods: A history of k-means algorithms. In *Selected Contributions in Data Analysis and Classification*, pages 161–172. Springer Berlin Heidelberg, 2007.

[29] Hans-Hermann Bock. Origins and extensions of the *k*-means algorithm in cluster analysis. *Electronic Journal for History of Probability and Statistics*, 4(2):1–18, 2008.

[30] Tossapon Boongoen and Natthakan Iam-On. Cluster ensembles: A survey of approaches with recent extensions and applications. *Computer Science Review*, 28:1–25, May 2018.

[31] Charles Bouveyron and Camille Brunet-Saumard. Model-based clustering of high-dimensional data: A review. *Computational Statistics and Data Analysis*, 71:52–78, 2014.

[32] Jianghui Cai, Jing Hao, Haifeng Yang, Xujun Zhao, and Yuqing Yang. A review on semi-supervised clustering. *Information Sciences*, 632:164–200, June 2023.

[33] R.J. Campello, E.R. Hruschka, and V.S. Alves. On the efficiency of evolutionary fuzzy clustering. *Journal of Heuristics*, 15(1):43–75, 2009.

[34] Y. Cao and J. Wu. Dynamics of projective adaptive resonance theory model: The foundation of part algorithm. *IEEE Transactions on Neural Networks*, 15(2):245–260, March 2004.

[35] Yongqiang Cao and Jianhong Wu. Projective ART for clustering data sets in high dimensional spaces. *Neural Networks*, 15(1):105–120, January 2002.

[36] Claude Cariou, Steven Le Moan, and Kacem Chehdi. A novel mean-shift algorithm for data clustering. *IEEE Access*, 10:14575–14585, 2022.

[37] J.W. Carmichael and R.S. Julius. Finding natural clusters. *Systematic Biology*, 17(2):144–150, June 1968.

[38] Gilles Celeux and Gérard Govaert. Gaussian parsimonious clustering models. *Pattern Recognition*, 28(5):781–793, May 1995.

[39] Jae-Woo Chang and Du-Seok Jin. A new cell-based clustering method for large, high-dimensional data in data mining applications. In *Proceedings of the 2002 ACM symposium on Applied computing*, SAC02, pages 503–507. ACM, March 2002.

[40] Guoqing Chao, Shiliang Sun, and Jinbo Bi. A survey on multiview clustering. *IEEE Transactions on Artificial Intelligence*, 2(2):146–168, 2021.

[41] Anil Chaturvedi, Paul E. Green, and J. Douglas Caroll. K-modes clustering. *Journal of Classification*, 18(1):35–55, January 2001.

[42] Chun-Hung Cheng, Ada Waichee Fu, and Yi Zhang. Entropy-based subspace clustering for mining numerical data. In *Proceedings of the fifth ACM SIGKDD international conference on Knowledge discovery and data mining*, KDD99. ACM, August 1999.

[43] Yizong Cheng. Mean shift, mode seeking, and clustering. *IEEE Transactions on Pattern Analysis and Machine Intelligence*, 17(8):790–799, 1995.

[44] D. Comaniciu and P. Meer. Mean shift analysis and applications. In *The Proceedings of the Seventh IEEE International Conference on ComputerVision*, volume 2, pages 1197–1203. IEEE, September 1999.

[45] D. Comaniciu and P. Meer. Mean shift: a robust approach toward feature space analysis. *IEEE Transactions on Pattern Analysis and Machine Intelligence*, 24(5):603–619, May 2002.

[46] R.M. Cormack. A review of classification. *Journal of the Royal Statistical Society. Series A (General)*, 134(3):321, 1971.

[47] D.R. Cox. Note on grouping. *Journal of the American Statistical Association*, 52(280):543–547, 1957.

[48] J.A. Cuesta-Albertos, A. Gordaliza, and C. Matrán. Trimmed k-means: an attempt to robustify quantizers. *The Annals of Statistics*, 25(2), April 1997.

[49] Zineb Dafir, Yasmine Lamari, and Said Chah Slaoui. A survey on parallel clustering algorithms for big data. *Artificial Intelligence Review*, 54(4):2411–2443, 2020.

[50] T. Dalenius and M. Gurney. The problem of optimum stratification. ii. *Scandinavian Actuarial Journal*, 1951(1–2):133–148, January 1951.

[51] Tore Dalenius. The problem of optimum stratification. *Scandinavian Actuarial Journal*, 1950(3–4):203–213, January 1950.

[52] Fatemeh Daneshfar, Sayvan Soleymanbaigi, Pedram Yamini, and Mohammad Sadra Amini. A survey on semi-supervised graph clustering. *Engineering Applications of Artificial Intelligence*, 133:108215, July 2024.

[53] Thi-Bich-Hanh Dao and Christel Vrain. A review on declarative approaches for constrained clustering. *International Journal of Approximate Reasoning*, 171:109135, August 2024.

[54] M. Dash, H. Liu, and X. Xu. '1+1>2': merging distance and density based clustering. In *Proceedings of the Seventh International Conference on Database Systemsfor Advanced Applications, 2001*, pages 32 –39, Hong Kong, China, April 2001.

[55] N.E. Day. Estimating the components of a mixture of normal distributions. *Biometrika*, 56(3):463–474, December 1969.

[56] Renato Cordeiro de Amorim. A survey on feature weighting based k-means algorithms. *Journal of Classification*, 33(2):210–242, July 2016.

[57] A.P. Dempster, N.M. Laird, and D.B. Rubin. Maximum likelihood from incomplete data via the EM algorithm. *Journal of the Royal Statistical Society. Series B (Methodological)*, 39(1):1–38, 1977.

[58] Zhaohong Deng, Kup-Sze Choi, Fu-Lai Chung, and Shitong Wang. Enhanced soft subspace clustering integrating within-cluster and between-cluster information. *Pattern Recognition*, 43(3):767–781, March 2010.

[59] Zhaohong Deng, Kup-Sze Choi, Yizhang Jiang, Jun Wang, and Shitong Wang. A survey on soft subspace clustering. *Information Sciences*, 348:84–106, June 2016.

[60] Inderjit S. Dhillon, Yuqiang Guan, and Brian Kulis. Kernel k-means: spectral clustering and normalized cuts. In *Proceedings of the tenth ACM SIGKDD international conference on Knowledge discovery and data mining*, KDD04. ACM, August 2004.

[61] C. Ding and H. Zha. *Spectral Clustering, Ordering and Ranking*. Springer-Verlag, New York, USA, 2010.

[62] Ling Ding, Chao Li, Di Jin, and Shifei Ding. Survey of spectral clustering based on graph theory. *Pattern Recognition*, 151:110366, July 2024.

[63] C. Döring, M.-J. Lesot, and R. Kruse. Data analysis with fuzzy clustering methods. *Computational Statistics and Data Analysis*, 51(1):192–214, 2006.

[64] J.L. DuBien and W.D. Warde. A mathematical comparison of the members of an infinite family of agglomerative clustering algorithms. *The Canadian Journal of Statistics*, 7:29–38, 1979.

[65] R. Duda, P. Hart, and D. Stork. *Pattern classification*. John Wiley & Sons, New York, NY, 2nd edition, 2001.

[66] J.C. Dunn. A fuzzy relative of the ISODATA process and its use in detecting compact well-separated clusters. *Journal of Cybernetics*, 3(3):32–57, January 1973.

[67] J.C. Dunn. Well-separated clusters and optimal fuzzy partitions. *Journal of Cybernetics*, 4(1):95–104, January 1974.

[68] A.W.F. Edwards and L.L. Cavalli-Sforza. A method for cluster analysis. *Biometrics*, 21(2):362–375, June 1965.

[69] Michael B. Eisen, Paul T. Spellman, Patrick O. Brown, and David Botstein. Cluster analysis and display of genome-wide expression patterns. *Proceedings of the National Academy of Sciences*, 95(25):14863–14868, December 1998.

[70] Y. El-Sonbaty, M.A. Ismail, and M. Farouk. An efficient density based clustering algorithm for large databases. In *16th IEEE International Conference on Tools with Artificial Intelligence, 2004. ICTAI 2004*, pages 673–677, November 2004.

[71] M. Ester, H. Kriegel, J. Sander, and X. Xu. A density-based algorithm for discovering clusters in large spatial databases with noise. In E. Simoudis, J. Han, and U. Fayyad, editors, *Second International Conference on Knowledge Discovery and Data Mining*, pages 226–231, Portland, Oregon, 1996. AAAI Press.

[72] B. Everitt, S. Landau, and M. Leese. *Cluster Analysis*. Oxford University Press, New York, fourth edition, 2001.

[73] B.S. Everitt. *Cluster Analysis*. Halsted Press, New York, Toronto, 3rd edition, 1993.

[74] Absalom E. Ezugwu, Abiodun M. Ikotun, Olaide O. Oyelade, Laith Abualigah, Jeffery O. Agushaka, Christopher I. Eke, and Andronicus A. Akinyelu. A comprehensive survey of clustering algorithms: State-of-the-art machine learning applications, taxonomy, challenges, and future research prospects. *Engineering Applications of Artificial Intelligence*, 110:104743, 2022.

[75] Absalom E. Ezugwu, Amit K. Shukla, Moyinoluwa B. Agbaje, Olaide N. Oyelade, Adán José-García, and Jeffery O. Agushaka. Automatic clustering algorithms: a systematic review and bibliometric analysis of relevant literature. *Neural Computing and Applications*, 33(11):6247–6306, 2020.

[76] V. Faber. Clustering and the continuous k-means algorithm. *Los Alamos Science*, 22:138–144, 1994.

[77] Adil Fahad, Najlaa Alshatri, Zahir Tari, Abdullah Alamri, Ibrahim Khalil, Albert Y. Zomaya, Sebti Foufou, and Abdelaziz Bouras. A survey of clustering algorithms for big data: Taxonomy and empirical analysis. *IEEE Transactions on Emerging Topics in Computing*, 2(3):267–279, 2014.

[78] Maurizio Filippone, Francesco Camastra, Francesco Masulli, and Stefano Rovetta. A survey of kernel and spectral methods for clustering. *Pattern Recognition*, 41(1):176–190, January 2008.

[79] R. A. Fisher. The use of multiple measurements in taxonomic problems. *Annals of Eugenics*, 7(2):179–188, September 1936.

[80] Walter D. Fisher. On grouping for maximum homogeneity. *Journal of the American Statistical Association*, 53(284):789–798, 1958.

[81] P. Foggia, G. Percannella, C. Sansone, and M. Vento. A graph-based clustering method and its applications. In *Proceedings of the 2nd international conference on Advances in brain, vision and artificial intelligence*, pages 277–287, Berlin, Heidelberg, 2007. Springer-Verlag.

[82] P. Foggia, G. Percannella, C. Sansone, and M. Vento. Benchmarking graph-based clustering algorithms. *Image and Vision Computing*, 27(7):979–988, June 2009.

[83] Edward Forgy. Cluster analysis of multivariate data: Efficiency vs. interpretability of classifications. *Biometrics*, 21(3):768–769, 1965.

[84] C. Fraley. How many clusters? which clustering method? answers via model-based cluster analysis. *The Computer Journal*, 41(8):578–588, August 1998.

[85] K. Fukunaga and L. Hostetler. The estimation of the gradient of a density function, with applications in pattern recognition. *IEEE Transactions on Information Theory*, 21(1):32–40, January 1975.

[86] Kunihiko Fukushima. Cognitron: A self-organizing multilayered neural network. *Biological Cybernetics*, 20(3–4):121–136, 1975.

[87] Y. Fukuyama and M. Sugeno. A new method of choosing the number of clusters for the fuzzy c-means method. In *Proceedings of 5th Fuzzy Syst. Symp.*, pages 247–250, 1989.

[88] Mohamed Medhat Gaber, Arkady Zaslavsky, and Shonali Krishnaswamy. Mining data streams: a review. *ACM SIGMOD Record*, 34(2):18–26, June 2005.

[89] G. Gan. *Subspace Clustering Based on Fuzzy Models and Mean Shifts*. PhD thesis, Department of Mathematics and Statistics, York University, Toronto, ON, Canada, March 2007.

[90] G. Gan, C. Ma, and J. Wu. *Data Clustering: Theory, Algorithms, and Applications*, volume 20 of *ASA-SIAM Series on Statistics and Applied Probability*. SIAM Press, SIAM, Philadelphia, ASA, Alexandria, VA, USA, 2007.

[91] G. Gan and J. Wu. A convergence theorem for the fuzzy subspace clustering (fsc) algorithm. *Pattern Recognition*, 41(6):1939–1947, June 2008.

[92] G. Gan, J. Wu, and Z. Yang. A fuzzy subspace algorithm for clustering high dimensional data. In X. Li, O.R. Zaiane, and Z. Li, editors, *Lecture Notes in Artificial Intelligence*, volume 4093, pages 271–278. Springer, August 2006.

[93] G. Gan, Z. Yang, and J. Wu. A genetic k-modes algorithm for clustering categorical data. In X. Li, S. Wang, and Z.Y. Dong, editors, *Proceedings on Advanced Data Mining and Applications: First InternationalConference, ADMA 2005, Wuhan, China*, volume 3584 of *Lecture Notes in Artificial Intelligence*, pages 195–202. Springer-Verlag GmbH, July 2005.

[94] Guojun Gan. *Data Clustering in C++: An Object-Oriented Approach*. Chapman & Hall/CRC Press, Boca Raton, FL, 2011.

[95] Guojun Gan and Michael Kwok-Po Ng. Subspace clustering using affinity propagation. *Pattern Recognition*, 48(4):1455–1464, April 2015.

[96] Guojun Gan and Jianhong Wu. Subspace clustering for high dimensional categorical data. *ACM SIGKDD Explorations Newsletter*, 6(2):87–94, December 2004.

[97] Guojun Gan, Jianhong Wu, and Zijiang Yang. Partcat: A subspace clustering algorithm for high dimensional categorical data. In *The 2006 IEEE International Joint Conference on Neural Network Proceedings*, pages 4406–4412. IEEE, 2006.

[98] Guojun Gan, Yuping Zhang, and Dipak K. Dey. Clustering by propagating probabilities between data points. *Applied Soft Computing*, 41:390–399, April 2016.

[99] V. Ganti, J. Gehrke, and R. Ramakrishnan. CACTUS: Clustering categorical data using summaries. In S. Chaudhuri and D. Madigan, editors, *Proceedings of the Fifth ACM SIGKDD International Conference on KnowledgeDiscovery and Data Mining*, pages 73–83, N.Y., August 1999. ACM Press.

[100] I. Gath and A.B. Geva. Unsupervised optimal fuzzy clustering. *IEEE Transactions on Pattern Analysis and Machine Intelligence*, 11(7):773–780, July 1989.

[101] Alan Genz and Frank Bretz. *Computation of Multivariate Normal and t Probabilities.* Springer, New York, NY, 2009.

[102] David Gibson, Jon Kleinberg, and Prabhakar Raghavan. Clustering categorical data: an approach based on dynamical systems. *The VLDB Journal The International Journal on Very Large Data Bases,* 8(3–4):222–236, February 2000.

[103] Fred Glover, Eric Taillard, and Eric Taillard. A user's guide to tabu search. *Annals of Operations Research,* 41(1):1–28, March 1993.

[104] S. Goil, H.S. Nagesh, and A. Choudhary. MAFIA: Efficient and scalable subspace clustering for very large datasets. Technical Report CPDC-TR-9906-010, Northwestern University, June 1999.

[105] Amin Golzari Oskouei, Negin Samadi, Shirin Khezri, Arezou Najafi Moghaddam, Hamidreza Babaei, Kiavash Hamini, Saghar Fath Nojavan, Asgarali Bouyer, and Bahman Arasteh. Feature-weighted fuzzy clustering methods: An experimental review. *Neurocomputing,* 619:129176, February 2025.

[106] A.D. Gordon. A review of hierarchical classification. *Journal of the Royal Statistical Society. Series A (General),* 150(2):119, 1987.

[107] A.D. Gordon. Hierarchical classification. In P. Arabie, L.J. Hubert, and G.De Soete, editors, *Clustering and Classification,* pages 65–121, River Edge, NJ, USA, 1996. World Scientific.

[108] Anjana Gosain and Sonika Dahiya. Performance analysis of various fuzzy clustering algorithms: A review. *Procedia Computer Science,* 79:100–111, 2016.

[109] J.C. Gower. A general coefficient of similarity and some of its properties. *Biometrics,* 27(4):857, December 1971.

[110] J.C. Gower and G.J.S. Ross. Minimum spanning trees and single linkage cluster analysis. *Applied Statistics,* 18(1):54, 1969.

[111] P. Grabusts and A. Borisov. Using grid-clustering methods in data classification. In *Proceedings. International Conference on Parallel Computing in Electrical Engineering,* PCEE-02, pages 425–426. IEEE Comput. Soc, 2022.

[112] S. Grossberg. Adaptive pattern classification and universal recoding: I. parallel development and coding of neural feature detectors. *Biological Cybernetics,* 23(3):121–134, 1976.

[113] Stephen Grossberg. Adaptive pattern classification and universal recoding: Ii. feedback, expectation, olfaction, illusions. *Biological Cybernetics,* 23(4):187–202, December 1976.

[114] M. Grötschel and Y. Wakabayashi. A cutting plane algorithm for a clustering problem. *Mathematical Programming*, 45(1-3):59–96, aug 1989.

[115] S. Guha, A. Meyerson, N. Mishra, R. Motwani, and L. O'Callaghan. Clustering data streams: theory and practice. *IEEE Transactions on Knowledge and Data Engineering*, 15(3):515–528, May 2003.

[116] S. Guha, R. Rastogi, and K. Shim. ROCK: a robust clustering algorithm for categorical attributes. In *Proceedings 15th International Conference on Data Engineering (Cat. No.99CB36337)*. IEEE, 1999.

[117] Sudipto Guha, Rajeev Rastogi, and Kyuseok Shim. CURE: an efficient clustering algorithm for large databases. In *Proceedings of the 1998 ACM SIGMOD international conference on Management of data*, SIGMOD/PODS98. ACM, June 1998.

[118] M. Halkidi, Y. Batistakis, and M. Vazirgiannis. Cluster validity methods: part I. *ACM SIGMOD Record*, 31(2), 2002.

[119] M. Halkidi, Y. Batistakis, and M. Vazirgiannis. Clustering validity checking methods: part II. *ACM SIGMOD Record*, 31(3), 2002.

[120] Emrah Hancer and Dervis Karaboga. A comprehensive survey of traditional, merge-split and evolutionary approaches proposed for determination of cluster number. *Swarm and Evolutionary Computation*, 32:49–67, February 2017.

[121] Emrah Hancer, Bing Xue, and Mengjie Zhang. A survey on feature selection approaches for clustering. *Artificial Intelligence Review*, 53(6):4519–4545, 2020.

[122] Muhammad Haris, Yusliza Yusoff, Azlan Mohd Zain, Abid Saeed Khattak, and Syed Fawad Hussain. Breaking down multi-view clustering: A comprehensive review of multi-view approaches for complex data structures. *Engineering Applications of Artificial Intelligence*, 132:107857, June 2024.

[123] Erez Hartuv and Ron Shamir. A clustering algorithm based on graph connectivity. *Information Processing Letters*, 76(4–6):175–181, December 2000.

[124] R. Hathaway, J.C. Bezdek, and W. Tucker. An improved covergence theorem for the fuzzy c-means clustering algorithms. In J.C. Bezdek, editor, *Analysis of Fuzzy Information*, volume III, pages 123–131. CRC Press, Inc., 1987.

[125] Richard J. Hathaway and James C. Bezdek. Local convergence of the fuzzy c-means algorithms. *Pattern Recognition*, 19(6):477–480, January 1986.

[126] Christian Hennig, Marina Meila, Fionn Murtagh, and Roberto Rocci, editors. *Handbook of cluster analysis.* Chapman & Hall/CRC Handbooks of modern statistical methods. CRC Press, Taylor & Francis Group, Boca Raton, FL, 2016.

[127] Rui Henriques and Sara C. Madeira. Triclustering algorithms for three-dimensional data analysis: A comprehensive survey. *ACM Computing Surveys*, 51(5):1–43, September 2018.

[128] J.H. Holland. *Adaptation in Natural and Artificial Systems.* University of Michigan Press, Ann Arbor, MI, 1975.

[129] Eric W. Holman. Statistical properties of large published classifications. *Journal of Classification*, 9(2):187–210, December 1992.

[130] F. Höppner, F. Klawonn, R. Kruse, and T. Runkler. *Fuzzy Cluster Analysis: Methods for Classification, Data Analysis and Image Recognition.* Wiley, 1999.

[131] Juhua Hu and Jian Pei. Subspace multi-clustering: a review. *Knowledge and Information Systems*, 56(2):257–284, October 2017.

[132] K.A. Hua, S.D. Lang, and W.K. Lee. A decomposition-based simulated annealing technique for data clustering. In ACM, editor, *Proceedings of the Thirteenth ACM SIGACT-SIGMOD-SIGART Symposium on Principles of Database Systems, May 24–26, 1994, Minneapolis, MN*, volume 13, pages 117–128, New York, NY 10036, USA, 1994. ACM Press.

[133] Huajuan Huang, Chen Wang, Xiuxi Wei, and Yongquan Zhou. Deep image clustering: A survey. *Neurocomputing*, 599:128101, September 2024.

[134] Z. Huang. Clustering large data sets with mixed numeric and categorical values. In *Knowledge discovery and data mining: techniques and applications.* World Scientific, 1997.

[135] Z. Huang. A fast clustering algorithm to cluster very large categorical data sets in data mining. In *SIGMOD Workshop on Research Issues on Data Mining and Knowledge Discovery*, Tucson, Arizona, May 1997.

[136] Zhexue Huang. Extensions to the k-means algorithm for clustering large data sets with categorical values. *Data Mining and Knowledge Discovery*, 2(3):283–304, 1998.

[137] Zhexue Huang and M.K. Ng. A fuzzy k-modes algorithm for clustering categorical data. *IEEE Transactions on Fuzzy Systems*, 7(4):446–452, 1999.

[138] M. Ichino. General metrics for mixed features the cartesian space theory for pattern recognition. In *Proceedings of the 1988 IEEE International Conference on Systems, Man, and Cybernetics*, volume 1, pages 494–497. IEEE, 1988.

[139] M. Ichino and H. Yaguchi. Generalized minkowski metrics for mixed feature-type data analysis. *IEEE Transactions on Systems, Man, and Cybernetics*, 24(4):698–708, April 1994.

[140] Abiodun M. Ikotun, Absalom E. Ezugwu, Laith Abualigah, Belal Abuhaija, and Jia Heming. K-means clustering algorithms: A comprehensive review, variants analysis, and advances in the era of big data. *Information Sciences*, 622:178–210, 2023.

[141] Abiodun M. Ikotun, Faustin Habyarimana, and Absalom E. Ezugwu. Cluster validity indices for automatic clustering: A comprehensive review. *Heliyon*, 11(2):e41953, January 2025.

[142] M.A. Ismail and Shokri Z. Selim. Fuzzy c-means: Optimality of solutions and effective termination of the algorithm. *Pattern Recognition*, 19(6):481–485, January 1986.

[143] Julien Jacques and Cristian Preda. Functional data clustering: a survey. *Advances in Data Analysis and Classification*, 8(3):231–255, November 2013.

[144] A.K. Jain, M.N. Murty, and P.J. Flynn. Data clustering: a review. *ACM Computing Surveys*, 31(3):264–323, September 1999.

[145] A.K. Jain. Data clustering: 50 years beyond *k*-means. *Pattern Recognition Letters*, 31(8):651–666, 2010.

[146] A.K. Jain and R.C. Dubes. *Algorithms for Clustering Data*. Prentice Hall,Englewood Cliffs, New Jersey, 1988.

[147] A.K. Jain, P.W. Duin, and Jianchang Mao. Statistical pattern recognition: a review. *IEEE Transactions on Pattern Analysis and Machine Intelligence*, 22(1):4–37, 2000.

[148] M. Jambu. *Classification automatique pour l'analyse de données*. Dunod, Paris, 1978.

[149] C.J. Jardine, N. Jardine, and R. Sibson. The structure and construction of taxonomic hierarchies. *Mathematical Biosciences*, 1(2):173–179, January 1967.

[150] Robert E. Jensen. A dynamic programming algorithm for cluster analysis. *Operations Research*, 17(6):1034–1057, dec 1969.

[151] Daxin Jiang, Chun Tang, and Aidong Zhang. Cluster analysis for gene expression data: a survey. *IEEE Transactions on Knowledge and Data Engineering*, 16(11):1370–1386, November 2004.

[152] Liping Jing, Michael K. Ng, and Joshua Zhexue Huang. An entropy weighting k-means algorithm for subspace clustering of high-dimensional sparse data. *IEEE Transactions on Knowledge and Data Engineering*, 19(8):1026–1041, August 2007.

[153] Stephen C. Johnson. Hierarchical clustering schemes. *Psychometrika*, 32(3):241–254, September 1967.

[154] Adán José-García and Wilfrido Gómez-Flores. Automatic clustering using nature-inspired metaheuristics: A survey. *Applied Soft Computing*, 41:192–213, April 2016.

[155] G. Karypis, Eui-Hong Han, and V. Kumar. Chameleon: hierarchical clustering using dynamic modeling. *Computer*, 32(8):68–75, 1999.

[156] L. Kaufman and P.J. Rousseeuw. *Finding Groups in Data–An Introduction to Cluster Analysis*. Wiley series in probability and mathematical statistics. John Wiley & Sons, Inc., New York, 1990.

[157] D. Keim and A. Hinneburg. Optimal grid-clustering: Towards breaking the curse of dimensionality inhigh-dimensional clustering. In *Proceedings of the 25th International Conference on Very Large Data Bases(VLDB '99)*, pages 506–517, San Francisco, September 1999. Morgan Kaufmann.

[158] Hans-Peter Kriegel, Peer Kröger, and Arthur Zimek. Clustering high-dimensional data: A survey on subspace clustering, pattern-based clustering, and correlation clustering. *ACM Transactions on Knowledge Discovery from Data*, 3(1):1–58, March 2009.

[159] K. Krishna and M. Narasimha Murty. Genetic k-means algorithm. *IEEE Transactions on Systems, Man and Cybernetics, Part B (Cybernetics)*, 29(3):433–439, June 1999.

[160] Nur Laila Ab Ghani, Izzatdin Abdul Aziz, and Said Jadid AbdulKadir. Subspace clustering in high-dimensional data streams: A systematic literature review. *Computers, Materials & Continua*, 75(2):4649–4668, 2023.

[161] G.N. Lance and W.T. Williams. A general theory of classificatory sorting strategies: 1. hierarchical systems. *The Computer Journal*, 9(4):373–380, February 1967.

[162] G.N. Lance and W.T. Williams. A general theory of classificatory sorting strategies: Ii. clustering systems. *The Computer Journal*, 10(3):271–277, March 1967.

[163] L. Legendre and P. Legendre. *Numerical Ecology.* Elsevier Scientific, New York, 1983.

[164] Yee Leung, Jiang-She Zhang, and Zong-Ben Xu. Clustering by scale-space filtering. *IEEE Transactions on Pattern Analysis and Machine Intelligence*, 22(12):1396–1410, 2000.

[165] M. Sh. Levin. Combinatorial clustering: Literature review, methods, examples. *Journal of Communications Technology and Electronics*, 60(12):1403–1428, December 2015.

[166] C. Li and G. Biswas. Unsupervised learning with mixed numeric and nominal data. *IEEE Transactions on Knowledge and Data Engineering*, 14(4):673–690, July 2002.

[167] Aristidis Likas, Nikos Vlassis, and Jakob J. Verbeek. The global k-means clustering algorithm. *Pattern Recognition*, 36(2):451–461, February 2003.

[168] N.P. Lin, C. Chang, H.-E. Chueh, H.-J. Chen, and W.-H. Hao. A deflected grid-based algorithm for clustering analysis. *WSEAS Transactions on Computers*, 7(4):125–132, 2008.

[169] Bing Liu, Yiyuan Xia, and Philip S. Yu. Clustering through decision tree construction. In *Proceedings of the ninth international conference on Information and knowledge management*, CIKM00, pages 20–29. ACM, November 2000.

[170] S.P. Lloyd. Least squares quantization in PCM. *IEEE Transactions on Information Theory*, 28(2):129 –137, 1982.

[171] M. Lorr. *Cluster Analysis for Social Scientists.* The Jossey-Bass Social and Behavioral Science Series. Jossey-Bass, San Francisco, Washington, London, 1983.

[172] Stephen Lynch. *A simple introduction to Python.* CRC Press, Boca Raton, FL, 2024.

[173] P. Macnaughton-Smith, W.T. Williams, M.B. Dale, and L.G. Mockett. Dissimilarity analysis: a new technique of hierarchical sub-division. *Nature*, 202(4936):1034–1035, June 1964.

[174] J.B. MacQueen. Some methods for classification and analysis of multi-variate observations. In L.M. LeCam and J. Neyman, editors, *Proceedings of the 5th Berkeley Symposium on Mathematical Statistics andProbability*, volume 1, pages 281–297, Berkely, CA, USA, 1967. University of California Press.

[175] Mahmoud A. Mahdi, Khalid M. Hosny, and Ibrahim Elhenawy. Scalable clustering algorithms for big data: A review. *IEEE Access*, 9:80015–80027, 2021.

[176] Jianchang Mao and A.K. Jain. A self-organizing network for hyper-ellipsoidal clustering (hec). *IEEE Transactions on Neural Networks*, 7(1):16–29, January 1996.

[177] W.L. Martinez and A.R. Martinez. *Exploratory data analysis with MAT-LAB*. Computer Science and Data Analysis. Chapman & Hall/CRC, Boca Raton, Florida, USA, 2005.

[178] G.J. McLachlan and T. Krishnan. *The EM Algorithm and Extensions*. Wiley, NY, USA, 1997.

[179] Paul D. McNicholas. Model-based clustering. *Journal of Classification*, 33(3):331–373, October 2016.

[180] X. Meng and D. van Dyk. The EM algorithm–an old folk-song sung to a fast new tune. *Journal of the Royal Statistical Society. Series B (Methodological)*, 59(3):511–567, 1997.

[181] Pierre Michaud. Clustering techniques. *Future Generation Computer Systems*, 13(2–3):135–147, November 1997.

[182] Glenn W. Milligan. Ultrametric hierarchical clustering algorithms. *Psychometrika*, 44(3):343–346, September 1979.

[183] Erxue Min, Xifeng Guo, Qiang Liu, Gen Zhang, Jianjing Cui, and Jun Long. A survey of clustering with deep learning: From the perspective of network architecture. *IEEE Access*, 6:39501–39514, 2018.

[184] S. Miyamoto, H. Ichihashi, and K. Honda. *Algorithms for Fuzzy Clustering: Methods in c-Means Clustering with Applications*. Springer-Verlag, Berlin, Heidelberg, 2008.

[185] N. Mladenović and P. Hansen. Variable neighborhood search. *Computers & Operations Research*, 24(11):1097–1100, November 1997.

[186] Donald G. Morrison. Measurement problems in cluster analysis. *Management Science*, 13(12):B-775–B-780, August 1967.

[187] Anirban Mukhopadhyay, Ujjwal Maulik, and Sanghamitra Bandyopadhyay. A survey of multiobjective evolutionary clustering. *ACM Computing Surveys*, 47(4):1–46, May 2015.

[188] John M. Mulvey and Harlan P. Crowder. Cluster analysis: An application of lagrangian relaxation. *Management Science*, 25(4):329–340, apr 1979.

[189] F. Murtagh. A survey of recent advances in hierarchical clustering algorithms. *The Computer Journal*, 26(4):354–359, November 1983.

[190] Fionn Murtagh. Counting dendrograms: A survey. *Discrete Applied Mathematics*, 7(2):191–199, February 1984.

[191] Emmanuel Müller, Stephan Günnemann, Ira Assent, and Thomas Seidl. Evaluating clustering in subspace projections of high dimensional data. *Proceedings of the VLDB Endowment*, 2(1):1270–1281, August 2009.

[192] H. Nagesh, S. Goil, and A. Choudhary. Adaptive grids for clustering massive data sets. In *First SIAM international conference on Data Mining*, Chicago, IL, USA, April 5-7 2001.

[193] Satyasai Jagannath Nanda and Ganapati Panda. A survey on nature inspired metaheuristic algorithms for partitional clustering. *Swarm and Evolutionary Computation*, 16:1–18, June 2014.

[194] Sami Naouali, Semeh Ben Salem, and Zied Chtourou. Clustering categorical data: A survey. *International Journal of Information Technology & Decision Making*, 19(01):49–96, January 2020.

[195] Mariá C.V. Nascimento and André C.P.L.F. de Carvalho. Spectral methods for graph clustering – a survey. *European Journal of Operational Research*, 211(2):221–231, June 2011.

[196] Andrew Y. Ng, Michael I. Jordan, and Yair Weiss. On spectral clustering: analysis and an algorithm. In *Proceedings of the 15th International Conference on Neural Information Processing Systems: Natural and Synthetic*, NIPS'01, page 849–856, Cambridge, MA, USA, 2001. MIT Press.

[197] Michael K. Ng and Joyce C. Wong. Clustering categorical data sets using tabu search techniques. *Pattern Recognition*, 35(12):2783–2790, December 2002.

[198] Hai-Long Nguyen, Yew-Kwong Woon, and Wee-Keong Ng. A survey on data stream clustering and classification. *Knowledge and Information Systems*, 45(3):535–569, 2014.

[199] Yigit Oktar and Mehmet Turkan. A review of sparsity-based clustering methods. *Signal Processing*, 148:20–30, July 2018.

[200] L. Orloci. An agglomerative method for classification of plant communities. *The Journal of Ecology*, 55(1):193, March 1967.

[201] Gbeminiyi John Oyewole and George Alex Thopil. Data clustering: application and trends. *Artificial Intelligence Review*, 56(7):6439–6475, 2022.

[202] Divya Pandove, Shivan Goel, and Rinkl Rani. Systematic review of clustering high-dimensional and large datasets. *ACM Transactions on Knowledge Discovery from Data*, 12(2):1–68, 2018.

[203] N.H. Park and W.S. Lee. Grid-based subspace clustering over data streams. In *Proceedings of the sixteenth ACM conference on Conference on information and knowledge management*, pages 801–810, New York, NY, USA, 2007. ACM.

[204] Lance Parsons, Ehtesham Haque, and Huan Liu. Subspace clustering for high dimensional data: a review. *ACM SIGKDD Explorations Newsletter*, 6(1):90–105, June 2004.

[205] A. Patrikainen and M. Meila. Comparing subspace clusterings. *IEEE Transactions on Knowledge and Data Engineering*, 18(7):902–916, July 2006.

[206] D. Pelleg and A. Moore. Accelerating exact k-means algorithms with geometric reasoning. In *Proceedings of the fifth ACM SIGKDD international conference on Knowledgediscovery and data mining*, pages 277–281, San Diego, California, United States, 1999. ACM Press.

[207] S.J. Phillips. Acceleration of k-means and related clustering algorithms. In D. Mount and C. Stein, editors, *ALENEX: International Workshop on Algorithm Engineering and Experimentation,LNCS*, volume 2409, pages 166–177, San Francicsco, CA, USA, 2002. Springer-Verlag Heidelberg.

[208] Cecilia M. Procopiuc, Michael Jones, Pankaj K. Agarwal, and T.M. Murali. A monte carlo algorithm for fast projective clustering. In *Proceedings of the 2002 ACM SIGMOD international conference on Management of data*, SIGMOD/PODS02, pages 418–427. ACM, June 2002.

[209] B.-Z. Qiu, X.-L. Li, and J.-Y. Shen. Grid-based clustering algorithm based on intersecting partition and density estimation. In *Proceedings of the 2007 international conference on Emerging technologies in knowledge discovery and data mining*, pages 368–377, Berlin, Heidelberg, 2007. Springer-Verlag.

[210] R. Michalski. Soybean (small), 1980.

[211] R. Quinlan. Auto MPG, 1993.

[212] M.R. Rao. Cluster analysis and mathematical programming. *Journal of the American Statistical Association*, 66(335):622–626, sep 1971.

[213] Sura Raya, Mariam Orabi, Imad Afyouni, and Zaher Al Aghbari. Multi-modal data clustering using deep learning: A systematic review. *Neurocomputing*, 607:128348, November 2024.

[214] Yazhou Ren, Jingyu Pu, Zhimeng Yang, Jie Xu, Guofeng Li, Xiaorong Pu, Philip S. Yu, and Lifang He. Deep clustering: A comprehensive survey. *IEEE Transactions on Neural Networks and Learning Systems*, pages 1–21, 2024.

[215] Hermes Robles-Berumen, Amelia Zafra, and Sebastián Ventura. A survey of genetic algorithms for clustering: Taxonomy and empirical analysis. *Swarm and Evolutionary Computation*, 91:101720, December 2024.

[216] F.J. Rohlf. Algorithm 81: Dendrogram plot. *The Computer Journal*, 17(1):89–91, February 1974.

[217] Lior Rokach. A survey of clustering algorithms. In *Data Mining and Knowledge Discovery Handbook*, pages 269–298. Springer US, 2009.

[218] Kenneth Rose, Eitan Gurewitz, and Geoffrey C. Fox. Statistical mechanics and phase transitions in clustering. *Physical Review Letters*, 65(8):945–948, August 1990.

[219] Samuel Rota Bulò and Marcello Pelillo. Dominant-set clustering: A review. *European Journal of Operational Research*, 262(1):1–13, October 2017.

[220] D.E. Rumelhart and D. Zipser. Feature discovery by competitive learning. In *Parallel distributed processing: explorations in the microstructure of cognition, vol. 1: foundations*, pages 151–193, Cambridge, MA, USA, 1986. MIT Press.

[221] E.H. Ruspini. A new approach to clustering. *Information and Control*, 15:22–32, 1969.

[222] Jörg Sander, Martin Ester, Hans-Peter Kriegel, and Xiaowei Xu. Density-based clustering in spatial databases: The algorithm GDB-SCAN and its applications. *Data Mining and Knowledge Discovery*, 2(2):169–194, 1998.

[223] Amit Saxena, Mukesh Prasad, Akshansh Gupta, Neha Bharill, Om Prakash Patel, Aruna Tiwari, Meng Joo Er, Weiping Ding, and Chin-Teng Lin. A review of clustering techniques and developments. *Neurocomputing*, 267:664–681, December 2017.

[224] E. Schikuta. Grid-clustering: an efficient hierarchical clustering method for very large data sets. In *Proceedings of 13th International Conference on Pattern Recognition*. IEEE, 1996.

[225] E. Schikuta and M. Erhart. The BANG-clustering system: Grid-based data analysis. In X. Liu, P. Cohen, and M. Berthold, editors, *Lecture Notes in Computer Science*, volume 1280, pages 513–524, Berlin Heidelberg, 1997. Springer-Verlag.

[226] J. Scoltock. A survey of the literature of cluster analysis. *The Computer Journal*, 25(1):130–134, February 1982.

[227] A.J. Scott and M.J. Symons. Clustering methods based on likelihood ratio criteria. *Biometrics*, 27(2):387–397, June 1971.

[228] George S Sebestyen. *Decision-making processes in pattern recognition*. Macmillan, New York, NY, 1962.

[229] Shokri Z. Selim and M.A. Ismail. K-means-type algorithms: A generalized convergence theorem and characterization of local optimality. *IEEE Transactions on Pattern Analysis and Machine Intelligence*, PAMI-6(1):81–87, January 1984.

[230] Gholamhosein Sheikholeslami, Surojit Chatterjee, and Aidong Zhang. Wavecluster: a wavelet-based clustering approach for spatial data in very large databases. *The VLDB Journal The International Journal on Very Large Data Bases*, 8(3–4):289–304, February 2000.

[231] R. Sibson. SLINK: An optimally efficient algorithm for the single-link cluster method. *The Computer Journal*, 16(1):30–34, January 1973.

[232] Jonathan A. Silva, Elaine R. Faria, Rodrigo C. Barros, Eduardo R. Hruschka, André C.P.L.F. de Carvalho, and João Gama. Data stream clustering: A survey. *ACM Computing Surveys*, 46(1):1–31, July 2013.

[233] Kelvin Sim, Vivekanand Gopalkrishnan, Arthur Zimek, and Gao Cong. A survey on enhanced subspace clustering. *Data Mining and Knowledge Discovery*, 26(2):332–397, February 2012.

[234] Jaswinder Singh and Damanpreet Singh. A comprehensive review of clustering techniques in artificial intelligence for knowledge discovery: Taxonomy, challenges, applications and future prospects. *Advanced Engineering Informatics*, 62:102799, October 2024.

[235] Tony Sintes. Sams teach yourself object oriented programming in 21 days, 2002. Includes index.

[236] Hugo Steinhaus. On the division of material bodies into parts. *Bulletin of the Polish Academy of Sciences*, Cl. III — Vol. IV(12):801–804, 1956. In French.

[237] Douglas Steinley. k-means clustering: A half-century synthesis. *British Journal of Mathematical and Statistical Psychology*, 59:1–34, 2006.

[238] Gero Szepannek. clustMixType: User-friendly clustering of mixed-type data in R. *The R Journal*, 10(2):200, 2019.

[239] Kamal Taha. Semi-supervised and un-supervised clustering: A review and experimental evaluation. *Information Systems*, 114:102178, March 2023.

[240] M. Teboulle. A unified continuous optimization framework for center-based clustering methods. *The Journal of Machine Learning Research*, 8:65–102, 2007.

[241] S. Theodoridis and K. Koutroubas. *Pattern Recognition*. Academic Press, London, 1999.

[242] Roberto Todeschini, Davide Ballabio, Veronica Termopoli, and Viviana Consonni. Extended multivariate comparison of 68 cluster validity indices. a review. *Chemometrics and Intelligent Laboratory Systems*, 251:105117, August 2024.

[243] J. Valente de Oliveira and W. Pedrycz. *Advances in Fuzzy Clustering and its Applications*. John Wiley & Sons, Inc., New York, NY, USA, 2007.

[244] C.J. van Rijsbergen. Algorithm 47: A clustering algorithm. *The Computer Journal*, 13(1):113–115, February 1970.

[245] Hrishikesh D. Vinod. Integer programming and the theory of grouping. *Journal of the American Statistical Association*, 64(326):506–519, jun 1969.

[246] Ulrike von Luxburg. A tutorial on spectral clustering. *Statistics and Computing*, 17(4):395–416, August 2007.

[247] Hong-Yu Wang, Jie-Sheng Wang, and Guan Wang. A survey of fuzzy clustering validity evaluation methods. *Information Sciences*, 618:270–297, December 2022.

[248] L. Wang and Z. Wang. CUBN: A clustering algorithm based on density and distance. In *2003 International Conference on Machine Learning and Cybernetics*, pages 108–112, November 2003.

[249] Pingxin Wang, Xibei Yang, Weiping Ding, Jianming Zhan, and Yiyu Yao. Three-way clustering: Foundations, survey and challenges. *Applied Soft Computing*, 151:111131, January 2024.

[250] W. Wang, J. Yang, and R.R. Muntz. STING: A statistical information grid approach to spatial data mining. In M. Jarke, M.J. Carey, K.R. Dittrich, F.H. Lochovsky, and P. Loucopoulosand M.A. Jeusfeld, editors, *Twenty-Third International Conference on Very Large Data Bases*, pages 186–195, Athens, Greece, August 1997. Morgan Kaufmann.

[251] Yizhang Wang, Jiaxin Qian, Muhammad Hassan, Xinyu Zhang, Tao Zhang, Chao Yang, Xingxing Zhou, and Fengjin Jia. Density peak clustering algorithms: A review on the decade 2014–2023. *Expert Systems with Applications*, 238:121860, March 2024.

[252] Joe H. Ward. Hierarchical grouping to optimize an objective function. *Journal of the American Statistical Association*, 58(301):236–244, March 1963.

[253] Joe H. Ward and Marion E. Hook. Application of an hierarchical grouping procedure to a problem of grouping profiles. *Educational and Psychological Measurement*, 23(1):69–81, April 1963.

[254] Peter Willett. Recent trends in hierarchic document clustering: A critical review. *Information Processing & Management*, 24(5):577–597, January 1988.

[255] D. Wishart. k-means clustering with outlier detection, mixed variables and missing values. In *Exploratory Data Analysis in Empirical Research*, pages 216–226. Springer Berlin Heidelberg, 2003.

[256] J.H. Wolfe. Pattern clustering by multivariate mixture analysis. *Multivariate Behavioral Research*, 5:329–350, 1970.

[257] Kyoung-Gu Woo, Jeong-Hoon Lee, Myoung-Ho Kim, and Yoon-Joon Lee. FINDIT: a fast and intelligent subspace clustering algorithm using dimension voting. *Information and Software Technology*, 46(4):255–271, March 2004.

[258] Xi Xiao, Hailong Ma, Guojun Gan, Qing Li, Bin Zhang, and Shutao Xia. Robust k-means-type clustering for noisy data. *IEEE Transactions on Neural Networks and Learning Systems*, pages 1–15, 2024.

[259] Dongkuan Xu and Yingjie Tian. A comprehensive survey of clustering algorithms. *Annals of Data Science*, 2(2):165–193, 2015.

[260] R. Xu and D.C. Wunsch, II. *Clustering*. Wiley-IEEE Press, Hoboken, New Jersey, 2009.

[261] R. Xu and D. WunschII. Survey of clustering algorithms. *IEEE Transactions on Neural Networks*, 16(3):645–678, May 2005.

[262] Rui Xu and Donald C. Wunsch. Clustering algorithms in biomedical research: A review. *IEEE Reviews in Biomedical Engineering*, 3:120–154, 2010.

[263] Xiaowei Xu, Jochen Jäger, and Hans-Peter Kriegel. A fast parallel clustering algorithm for large spatial databases. *Data Mining and Knowledge Discovery*, 3(3):263–290, 1999.

[264] Jiong Yang, Wei Wang, Haixun Wang, and P. Yu. δ-clusters: capturing subspace correlation in a large data set. In *Proceedings 18th International Conference on Data Engineering*, ICDE-02, pages 517–528. IEEE Comput. Soc, 2002.

[265] M.-S. Yang. A survey of fuzzy clustering. *Mathematical and Computer Modelling*, 18(11):1–16, December 1993.

[266] Yan Yang and Hao Wang. Multi-view clustering: A survey. *Big Data Mining and Analytics*, 1(2):83–107, June 2018.

[267] Guoxian Yu, Liangrui Ren, Jun Wang, Carlotta Domeniconi, and Xiangliang Zhang. Multiple clusterings: Recent advances and perspectives. *Computer Science Review*, 52:100621, May 2024.

[268] Heng Yu and Xiaolan Hou. Hierarchical clustering in astronomy. *Astronomy and Computing*, 41:100662, October 2022.

[269] L.A. Zadeh. Fuzzy sets. *Information and Control*, 8:338–353, 1965.

[270] O.R. Zaiane and C. Lee. Clustering spatial data in the presence of obstacles: a density-based approach. In *Proceedings. International Database Engineering and Applications Symposium,2002.*, pages 214–223, July 2002.

[271] B. Zhang, M. Hsu, and U. Dayal. k-harmonic means - a spatial clustering algorithm with boosting. In *Proceedings of the First International Workshop on Temporal, Spatial,and Spatio-Temporal Data Mining-Revised Papers*, volume 2007 of *Lecture Notes in Computer Science*, pages 31–45, London, UK, 2001. Springer-Verlag.

[272] Bin Zhang. *Comparison of the Performance of Center-Based Clustering Algorithms*, pages 63–74. Springer Berlin Heidelberg, 2003.

[273] Tian Zhang, Raghu Ramakrishnan, and Miron Livny. BIRCH: an efficient data clustering method for very large databases. *ACM SIGMOD Record*, 25(2):103–114, June 1996.

[274] Zuowei Zhang, Yiru Zhang, Hongpeng Tian, Arnaud Martin, Zhunga Liu, and Weiping Ding. A survey of evidential clustering: Definitions, methods, and applications. *Information Fusion*, 115:102736, March 2025.

[275] Y. Zhao and J. Song. GDILC: a grid-based density-isoline clustering algorithm. In *International Conferences on Info-tech and Info-net, 2001. Proceedings.ICII 2001*, volume 3, pages 140–145, Beijing, China, November 2001. IEEE.

[276] S. Zhong and J. Ghosh. A unified framework for model-based clustering. *The Journal of Machine Learning Research*, 4:1001–1037, 2003.

Index

For Product Safety Concerns and Information please contact our EU
representative GPSR@taylorandfrancis.com
Taylor & Francis Verlag GmbH, Kaufingerstraße 24, 80331 München, Germany

www.ingramcontent.com/pod-product-compliance
Lightning Source LLC
Chambersburg PA
CBHW050525190326
41458CB00005B/1662

* 9 7 8 1 0 3 2 9 7 1 5 6 8 *